Berliner ökophysiologische

und phytomedizinische Schriften

Hrsg. von Christian Ulrichs und Carmen Büttner

Lebenswissenschaftliche Fakultät,

Humboldt-Universität zu Berlin

Band 49

Hrsg. von

Prof. Dr. Dr. Christian Ulrichs
Humboldt-Universität zu Berlin

und

Dr. Inga Mewis
Julius Kühn-Institut, Berlin
Humboldt-Universität zu Berlin

Chemical ecology of Cabbage White (*Pieris* sp.)

Dissertation

Zur Erlangung des akademischen Grades

Doctor rerum agriculturarum

(Dr. rer. agr.)

Im Fach Agrarwissenchaft

eingereicht an der

Lebenswissenschaftlichen Fakultät

der Humboldt-Universität zu Berlin

von

Maliha Gul Aftab

Geboren am 11.08.1982

Präsidentin der Humboldt-Universität zu Berlin

Prof. Dr.-Ing. Dr. Sabine Kunst

Dekan der Lebenswissenschaftliche Fakultät

Prof. Dr. Dr. Christian Ulrichs

Gutachter/innen:

1. Prof. Dr. Dr. Christian Ulrichs
2. Prof. Dr. Stefan Kühne
3. Dr. Carsten Müller

Tag der mündlichen Prüfung: 01.02.2021

Bibliografische Information der Deutschen Nationalbibliothek

Die Deutsche Nationalbibliothek verzeichnet diese Publikation in der Deutschen Nationalbibliografie; detaillierte bibliographische Daten sind im Internet über http://dnb.d-nb.de abrufbar.

1. Aufl. - Göttingen: Cuvillier, 2021

Zugl.: Berlin, Univ., Diss., 2021

Nonnenstieg 8, 37075 Göttingen

Telefon: 0551-54724-0

Telefax: 0551-54724-21

www.cuvillier.de

1. Auflage, 2021

Gedruckt auf umweltfreundlichem, säurefreiem Papier aus nachhaltiger Forstwirtschaft.

ISBN 978-3-7369-7498-2

eISBN 978-3-7369-6498-3

Dedicated to my Family

Table of Contents

1. General introduction and aim of the thesis

Chemical ecology explores the structure, origin, and function of naturally occurring chemicals that mediate intraspecific or interspecific interactions. These chemicals are known as semiochemicals. Depending on their function, they were recently divided into three major classes: pheromones, kairomones, and allomones (Drijfhout 2017). Wyatt explained pheromones as "molecules that are evolved signals, in defined ratios in the case of multiple component pheromones, which are emitted by an individual and received by a second individual of the same species, in which they cause a specific reaction, for example, a stereotyped behavior or a developmental process" (Wyatt 2010; modified after Karlson and Lüscher 1959).

Among Lepidoptera insects, butterflies are about 9% of the order worldwide, totaling about 20,400 described species. They are one of the most widespread and widely recognized insects (Heppner 2008; Powell and Opler 2009). They are currently grouped within the taxonomic superfamily Papilionoidea, with seven families, namely Papilionidae, Pieridae, Riodinidae, Lycaenidae, Nymphalidae, Hesperiidae (skippers), and Hedylidae (Van Nieukerken et al. 2011). In insects like Lepidoptera, the existence of communication via pheromone is widespread and widely discussed. Many different types of pheromones have been identified such as sex pheromones (Boeckh and Ernst 1987; Kanzaki et al. 1992; Hildebrand and Shepherd 1997) and social pheromones (Hölldobler and Wilson 1990; Mizunami et al. 2010; Verheggen et al. 2010).

Butterflies are day-flying insects (excluding hedylids) whose partner-finding strategy is mainly based on visual and olfactory cues. The female butterflies do not have sex pheromone glands in their ovipositor, unlike moths. Therefore, they do not release long-range pheromones to find their partners. Instead, the male butterflies used their vision to detect conspecific females at some distance and to pursue them. The males release short-range pheromones to attract females, resulting in courtship (Silberglied 1984; Monteys et al. 2016). These short-range pheromones produced in the wing's scales are the bouquet of males, which help females in selecting their conspecifics before courtship.

The structure of wings' scales called androconia was found in Pierid species such as *P. brassicae*, *P. rapae*, and *P. napi*. Each Pierid species has characteristic pheromones produced in these androconia (Bergström and Lundgren 1973). In *P. rapae*, olfactory cues in recognition of partners and courtship behavior were studied and concluded that males could be attracted to females enclosed in Petri dishes (Obara 1964; Obara and Hidaka 1968). After visual cues, the smell was also reputed to play a part in courtship and attraction of imagines at close quarters to bring insects together (Hertz 1927). The compounds produced in male wings of *P. rapae* and *P. brassicae*, known as aphrodisiac pheromone, have been identified (Yildizhan et al. 2009). These are ferrulactone, hexahydrofarnesylacetone, and E-phytol in *P. rapae*, while the presence of brassicalactone instead of ferrulactone in *P. brassicae* has been identified (figure 1.1).

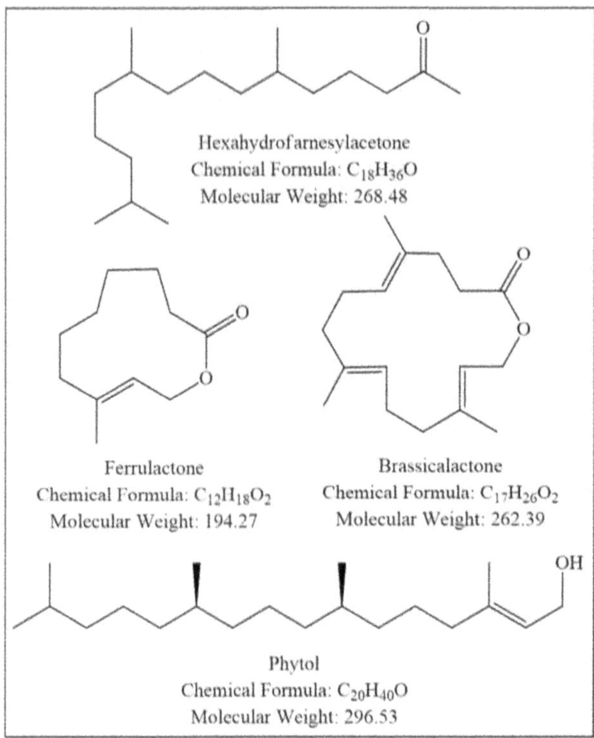

Figure 1.1: Chemical Structures of pheromones of wings of the *Pieris* species (Yildizhan et al. 2009)

To ensure environment-friendly agricultural practices, chemicals and/or pesticides must be avoided. Therefore, it becomes essential to find modern environmentally safe methods to affect cabbage white pests. Biological control of Lepidoptera has traditionally relied on pathogens, such as viruses (Harcourt 1966; Biever and Wilkinson 1978) and nematodes (Peters 1996). Insect natural enemies such as predators and parasitoids have also been used (Lucas et al. 1998; van Driesche and Hoddle 1997; Harvey et al. 1999; Karowe and Schoonhoven 1992). Another important way of control can be found out by studying the chemical ecology of Lepidoptera. The components of the male sex pheromone of the African butterfly *Bicyclus anynana* (Butler) have been explored to be (Z)-9-tetradecenol (Z9-14: OH), hexadecanal (16: Ald), and 6,10,14-

trimethylpentadecan-2-ol (6,10,14-trime-15-2-ol). Male sex pheromones of queen butterfly, *Danaus gilippus* (Berenice), and a butterfly in the subfamily Danainae, *Idea leuconoe* (Erichson), have also been identified (Pliske and Eisner 1969; Nishida et al. 1996). Male sex pheromone of *Pieris napi* has also been reported about a decade ago (Andersson et al. 2007). One way to control the cabbage butterflies is using other IPM strategies such as pheromone traps and mass trapping, where the studies regarding insect pheromones are considered particularly important. Recently Mating disruption of *Cossus insularis* (Staudinger) (Lepidoptera: Cossidae) with a synthetic pheromone was tested and resulted in a significantly decreased percentage of damaged trees (Hoshi et al. 2016).

The second aspect of chemical ecology also deals with the study of defense compounds produced by host plants. Many *Brassica* species are considered host plants of *Pieris* species (Feltwell 1982) that produce secondary metabolites known as glucosinolates (GS). They have a characteristic chemical structure with a sulfur-linked ß-D-glucopyranose moiety and an amino acid-derived side chain (figure 1.2). They are further divided based on the side chain into three groups, aliphatic, aromatic, and indole GS. A very comprehensive compilation of structures, trivial names, and distribution has been reviewed in which more than 140 GS have been listed (Fahey et al. 2001; European food safety authority 2008). GS-containing plants possess GS hydrolyzing enzyme, myrosinase, that is spatially separated from its substrate in the intact plant (Kelly et al. 1998). GS does not seem to be toxic themselves, but when they are brought together with myrosinase, they are rapidly hydrolyzed to toxic isothiocyanates (Cole 1976; Daxenbichler et al. 1977; Hanschen and Schreiner 2017) and some other compounds such as nitriles and epithionitriles (Halkier and Gershenzon 2006; Fenwick et al. 1983).

GS derivatives have a versatile range of functions. Some of them have a protective effect against cancer. Some GS derivatives may have detrimental effects, such as progoitrin and some derived from alkenyl GS. These compounds are mostly reported in rapeseed meals, which act as antinutrients affecting animal growth and development and lowering food intake (Rosa et al. 1997). For this study, four aliphatic and four indolic GS were chosen. Aliphatic GS includes 2-Hydroxy-3-butenyl, 2-Propenyl, 4-Methylsulfinylbutyl, and 3- Butenyl, while the indolic GS includes 4-Hydroxyindol-3-ylmethyl, Indol-3-ylmethyl, 4-Methoxyindol-3-ylmethyl, and 1-Methoxyindol-3-ylmethyl GS (figure 1.3).

P. rapae and *P. brassicae* have been observed as oligophagous insects feeding upon a wide range of plants. They can feed on about 105 plant families. A total of 83 species of families, Brassicaceae,

Papilionaceae, Resedaceae, Capparaceae, and Tropaeolaceae, are considered principal families exploited by *Pieris* species (Feltwell 1982; Bhandari et al. 2009). The close relation between herbivorous insects and their host plants has been widely discussed for decades. Prof. L.M. Schoonhoven's research describes that this relationship is due to a highly specialized sensory system of the larva and imago and their responses to chemical substances (GS) produced by the plants (Schoonhoven 1967; 1969). The caterpillars of *Pieris* are specialist herbivores that feed only on plants in the order Capparales that produce GS (Harvey et al. 2010). The adult females and larvae of *P. rapae* and *P. brassicae* use GS in Brassica plants for host selection for oviposition and feeding (Renwick et al. 1992; van Loon et al. 1992; Moyes et al. 2000; Schoonhoven and van Loon 2002).

The feeding of *P. brassicae* and *P. rapae* larvae upon different parts such as flowers and leaves of *Brassica nigra* var. *abyssinica* A. Braun was studied. As a result of larval feeding, allyl GS (sinigrin) levels become higher than in undamaged plants (Smallegange et al. 2007). Aphids increased aliphatic GS content in *Arabidopsis thaliana,* but *P. rapae* did not cause a significant induction of aliphatic GS, while the indolyl GS content increased by about 20% (Mewis et al. 2006).

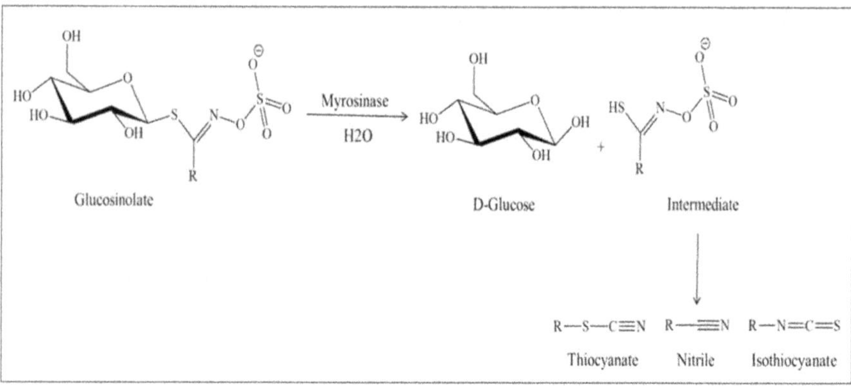

Figure 1.2: Products of GS hydrolysis

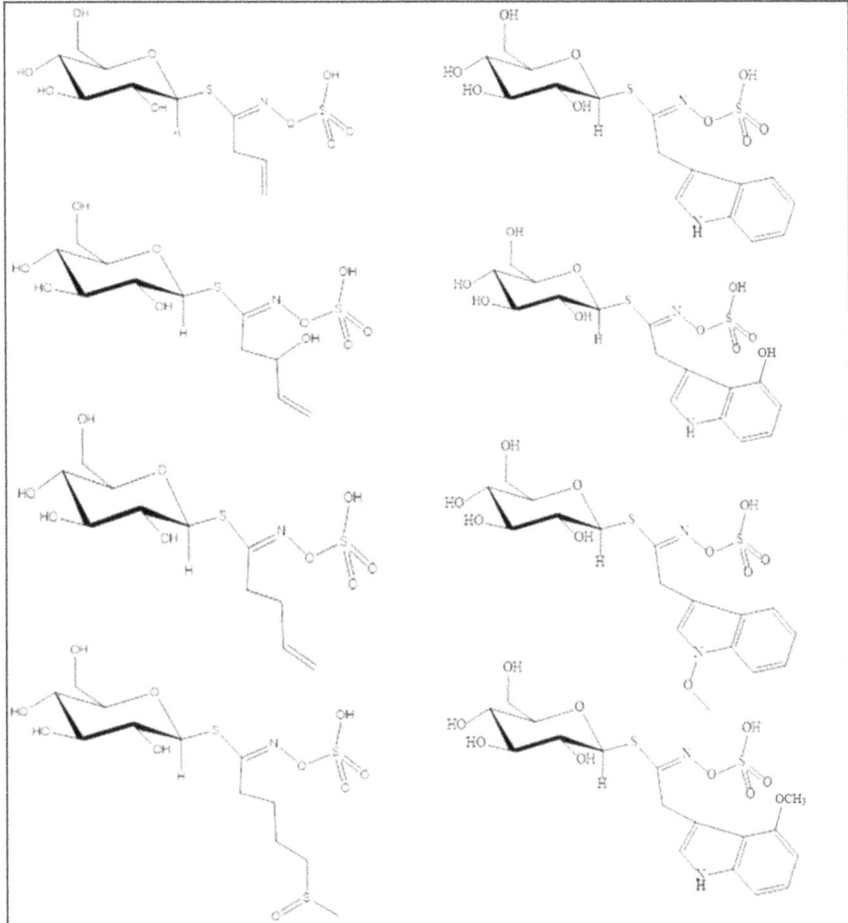

Figure 1.3: Structures of aliphatic and indolic GS considered in this study. Left upside down: 2-propenyl, 2-hydroxy-3-butenyl, 3-Butenyl, and 4-Methylsulfinylbutyl-GS. Right upside down: indol-3-ylmethyl, 4-Hydroxyindol-3-ylmethyl, 1-methoxyindol-3-ylmethyl GS, and 4-methoxyindol-3-ylmethyl-GS.

Short Introduction to Pieris species – The two most destructive pest species within the genus *Pieris* are *Pieris rapae* L. (small cabbage white) and *Pieris brassicae* L. (large cabbage white). Both species are native and widespread in Eurasia but have also been introduced to other continents where they have become well established. According to the datasheet published by the Centre for Agriculture and bioscience international (CABI.org), *P. rapae* is considered a pest in South America, Canada, Europe, Asia, northern countries of Africa, and Australia (figure 1.4). At the same time, *P. brassicae* is considered a pest mainly in the subtropical region of Asia and Africa and the whole European region (figure 1.4).

P. rapae was reported as one of the most destructive pests of Cole crops throughout the United States of America too (Kirby and Slosser 1984). According to a study in the western USA, *Pieris* species have caused about 41% of annual crucifers' losses (Shapiro 1975). It was also considered to be one of the significant defoliators of broccoli crops that led to major market losses (Vial et al. 1991; Cranshaw and Default 1985). Without pest management strategies, *P.rapae* larvae can cause complete crop loss in crucifers (Hely et al. 1982).

About 37 pest species of cruciferous plants have been listed from India, whereby *P. rapae* and *P. brassicae* were considered to be the most destructive ones (Lal 1975). According to a report issued in 2007, Indian agriculture has been suffered from yield losses by pests that can lead to 60-70% of yield loss (Dicke et al. 2007). *P. brassicae* is widely distributed and seen in all parts of India and Pakistan where Cole crops are widely cultivated. Larvae cause damage to all growth stages of plants (Lal and Bhajan 2004; Khan et al. 2017; Sachen and Gangwar 1980). According to a study carried out in Pakistan's northern areas, *P. brassicae* caused severe damages to all the *Brassica* cultivars used for the study. The highest population recorded was about 86 larvae per plant in the infestation peak season (Hasan and Ansari 2011; Yonus et al. 2004).

Life Cycle of *P. rapae* and *P. brassicae* - The life cycle of both species of cabbage butterflies has some differences, as shown in figures 1.5 and 1.6. *P. rapae* lay their eggs singly while *P. brassicae* lays 50-100 eggs in batches. *P. brassicae* shares the characteristic of laying its ova in batches with eight other British butterflies (Ford 1920; 1945). By laying single eggs, *P. rapae* has the advantage of being able to exploit isolated plants, and therefore it can be less discriminatory when choosing a suitable host for oviposition. A positive correlation between the number of eggs laid and the plant size has been recognized for *P. rapae*, predicting the minimization of larval food competition (Jones 1977, Ives 1978). *P. brassicae* is adapted for laying egg clusters in clumped vegetation and is able to maximize the utility of the host plants where larval migration does not constitute an inordinately high mortality risk (Davies and Gilbert 1985).

Both *Pieris* species select only leaves of cruciferous plants containing mustard oil glycosides on which to oviposit (Hewitt 1917). When offered cabbage and Reseda, *P. brassicae* females oviposit on cabbage first, and when the cabbage is overloaded, they switched to Reseda plants (Rothschild and Schoonhoven 1977). As the orange-yellow-colored ova mature, their color changed to a darker shade. After the larva emerges, it eats the egg case first, which contains vitamin A precursors required for growth (Fabre 1912). Both *Pieris* species have five instars and go through four larval-larval ecdyses (Klein 1932; Gardiner 1978). The larvae of *P. rapae* are velveted green with head setae partly black in color, whereas the larvae of *P. brassicae* are mottled, green with a black-colored pattern. *P. brassicae* larva is agiler than *P. rapae* over obstacles, and pupae can be found in high situations (Richards 1940). The larvae start to pupate only when they have enough materials left to make the silk girdle and cremaster (Eliot 1944). *P. brassicae* pupa is quite distinct by its large size and green, black and yellow mottling while *P. rapae* pupa is comparatively small and fresh green colored. *P. brassicae* pupae clearly show sexual differences. The female has two genital openings and is much broader than the males, as shown in figure 1.7 (Jackson 1890; Dusaussoy and Delplanque 1964). Precopulation courtship in *Pieris* species occurs in the heat periods of the day (Avidov and Harpaz 1969). Copulation in *P. brassicae* commences with the male landing on or beside the female. Then the male butterfly curls its abdomen across so that it approaches the female genitalia from below, and she folded wings and making a union. Afterward, the male turns around to face away from the female (David and Gardiner 1961; Mukherji 1961). Their mating lasts from 2 to 4 hours (Chandra and Lal 1977b).

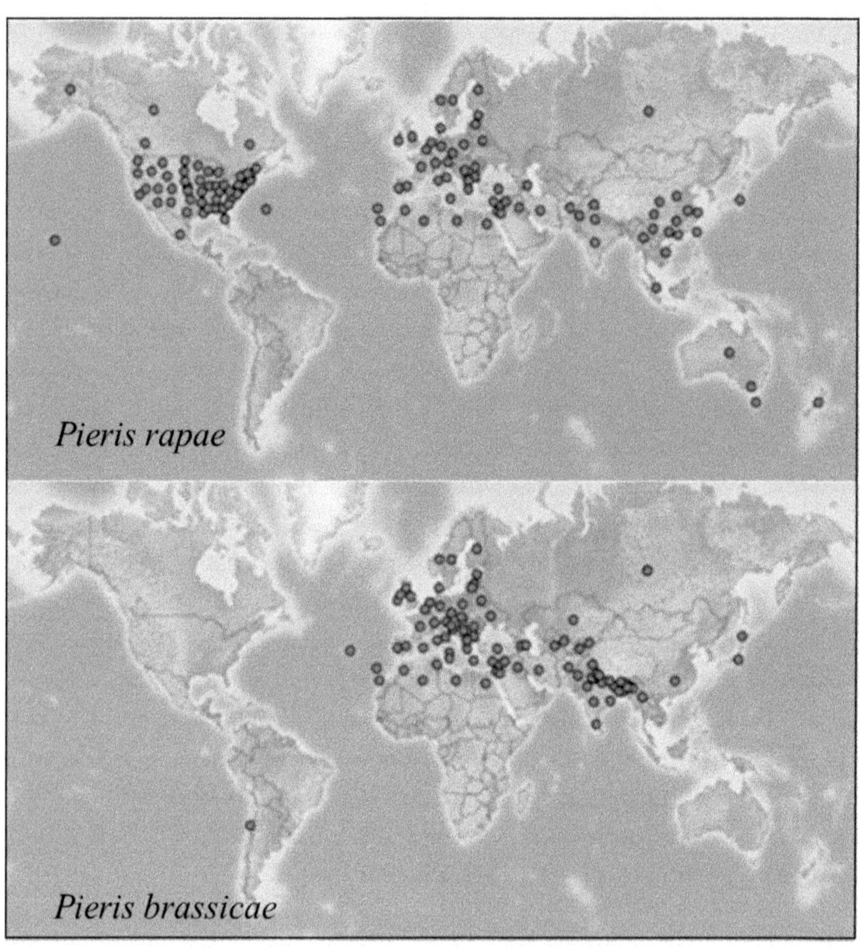

Figure 1.4: Maps showing the global distribution of *Pieris* species (CABI, Centre of agriculture and bioscience international; 2018)

Figure 1.5: Life cycle of *Pieris brassicae*. Adult butterflies (A); Eggs laid in a group (B); Third-star larvae feeding on a Turnip cabbage plant (C); Pupae (D)

Figure 1.6: Life cycle of *Pieris rapae*. Adult butterfly (A); Eggs laid separately on a leaf (B); Third-star larvae feeding on Turnip cabbage plants (C); Pupae (D)

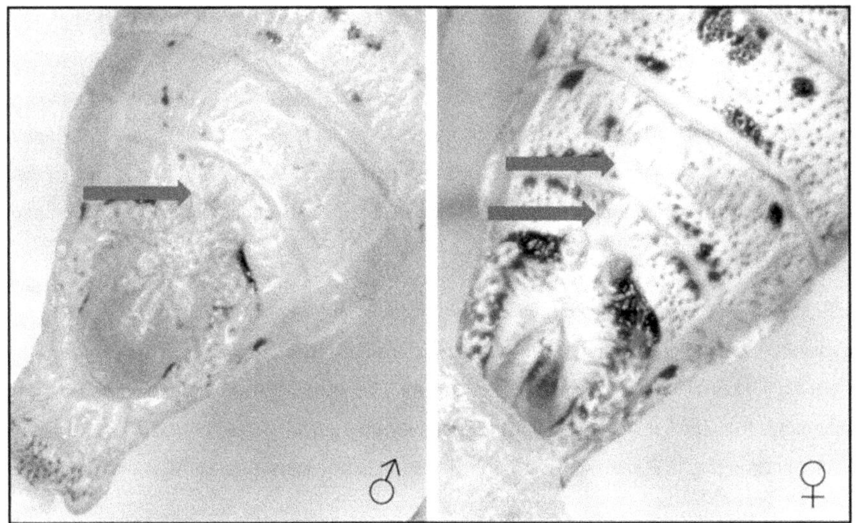

Figure 1.7: Sexual differences of pupae of *P. brassicae*. Male pupae with only one opening called aperture of ductus ejaculatoris (left); Female pupae with two openings named as aperture of bursa coplatrix and aperture of oviduct(right)

Aim of the thesis - The chemical ecology of an herbivore deals with the study of semiochemicals, which mediate insect-insect relationship (pheromones) as well as insect-host plant relationship (allelochemicals) (Nordlund et al. 1981). As the concept of integrated pest management is getting famous, these semiochemicals have been widely studied and used to control agricultural pest insects. In this regard, sex pheromones of moths and aphrodisiac, as well as anti-aphrodisiac pheromones of various butterflies, have been deeply studied. A synthetic sex pheromone of Carpenter moth, *Cossus insularis* (Staudinger), have been identified and tested to control this pest in Apple orchards via mating disruption strategy (Chen et al. 2006; Hoshi et al. 2016). In a green-veined butterfly, *Pieris napi*, a male produces anti-aphrodisiac pheromone methyl salicylate, which is transferred to females during mating, which renders females unattractive to other males (Andersson et al. 2000). Aphrodisiac pheromones found in the male wings of two species of sulfur butterflies, *Colias eurytheme* and *Colias philodice*, have been identified. It was found that male *C. philodice* produce three n-hexyl esters; myristate, palmitate, and stearate, whereas the male of *C. eurytheme* produce 13-methylheptacosane. These chemicals play an essential role in specie-recognition. Hence, it proves that aphrodisiac pheromones produced in the wings of male butterflies are specie- specific.

Aphrodisiac pheromones of *P. rapae* and *P. brassicae* have already been studied. The insects used for that experiment were from Netherlands (Yildizhan et al. 2009). According to a study about pheromones of *Choristoneura rosaceana* Harris, it was explored that the male moth's response to female sex pheromones in a field trapping experiment were location-dependent. Only 1% addition of (Z)-11-tetradecen-1-ol to a specific compound blend resulted in a twofold increase in average trap catch in British Columbia, Ontario, and Quebec, but this compound did not affect the insect population of Michigan or New York (El-syed et al. 2003).

With this background, this thesis aimed to investigate the male-specific aphrodisiac pheromones of two diverse populations of each subspecies of cabbage white butterfly, that are *P. rapae rapae* (Germany, Netherlands) and *P. rapae crucivora* (Taiwan, Vietnam). Some specific wing compounds from all Asian and European populations were quantified and compared in chapter II. Electroantennographical analyses (EAG) of *P. rapae* were also carried out and discussed in chapter III.

Chapter IV's focus was to identify and quantify the specific wings compounds of *P. brassicae* collected from trap fields in Berlin, Germany. EAG studies were also established, and a behavioral bioassay was carried out to investigate the response of female butterflies to the male wings extract and some synthetic compounds. These synthetic compounds were checked so that they can be used as pheromone lures in the field trials.

Another aspect of chemical ecology is the study of secondary metabolites of plants, known as Glucosinolates (GS). The adult females and larvae of *P. rapae* and *P. brassicae* use GS in *Brassica* plants for host selection for oviposition and feeding (Renwick et al. 1992; van Loon et al. 1992; Moyes et al. 2000; Schoonhoven and van Loon 2002). According to one study on *Psylliodes chrysocephala* L., it is observed that upon herbivory on *B. napus*, the aliphatic GS levels were decreased, while indolyl GS were increased and by mechanical wounding too, the same result was observed (Koritsas et al. 1989; 1991). In chapter V, we studied the herbivore-host plant relationship, where we discussed the host plant suitability of some brassica cultivars for *P. rapae* and *P. brassicae* and plants' response as a result of herbivory.

2 Chemical profiling of male-specific Aphrodisiac pheromones of *Pieris rapae* populations from different geographic regions

2.1 Abstract

Pieris rapae L. (small cabbage white butterfly) is a worldwide pest on crucifers. It is known that the species produce male-specific wings volatiles, also known as pheromones. This study investigated the pheromones variability found in the wings of *P. rapae* populations from different geographic origins. Two Asian populations, the race *P. rapae crucivora* (Taiwan, Vietnam) was compared with two European populations, the race *P. rapae rapae* (Germany, Netherlands). Using GC-MS-TOF and GC-FID, three main wings volatiles were identified: ferrulactone, hexahydrofarnesylacetone, and E-phytol. Ferrulactone was the male-specific component present in all studied *P. rapae* populations. E-Phytol was a part of male-specific pheromones blend in all populations except the population from the Netherlands. Hexahydrofarnesylacetone was present in wings of males and females but usually in higher amounts in males. Overall, the pheromone levels in the European populations were lower than the amounts present in Asian populations. In the populations from Europe, ferrulactone was one of the major constituents of male-specific pheromones.

Interestingly E-phytol was the major pheromone compound in Asian populations from Taiwan and Vietnam, with the relatively same ratios of the other two components. Therefore, both the subspecies of *P. rapae* show quantitatively a distinct wings pheromone profile. Comparing all four studied populations, it was concluded that they have the divergence in wings profile either. Thereby, our study can play an essential role in understanding reproductive isolation among diverse populations.

2.2 Introduction

Insects from the order Lepidoptera produce various pheromones to trigger behavioral responses (Karlson and Betenandt 1959; Karlson and Schneider 1973; Andersson et al. 2007; Witzgall et al. 2010). For example, *Danaus gilippus* (Cramer) produces pheromones during courtship behavior to facilitate mating (Meinwald et al. 1966). Males of *Eurema lisa* (Boisduval & Leconte) and *Colias philodice* (Godart) produce epicuticular components that facilitate females' acceptance behavior (Rutowski 1977a; 1977b; 1980; 1991). Some volatile compounds are known, which are transferred from males to females during courtship, known as anti-aphrodisiacs. In *Pieris napi* (L.), methyl salicylate, and in *Pieris brassicae* L., benzyl cyanide are transferred. Thus, making the females chemical mimics of males (Andersson et al. 2003). However, aphrodisiac pheromones are released by butterflies to attract their conspecifics. Some males use a mixture of various compounds in a specific proportion to attract females before courtship. In some Lepidoptera species, male-specific compounds are released from wings to stimulate female butterflies for mating. The aphrodisiac compounds are released from so-called scent scales located in the anteroposterior direction of male wings (Yoshida et al. 2000). The small cabbage white butterfly, *Pieris rapae* L., is a common pest of brassicaceous plants. This species causes severe destruction all around the world, such as in eastern Canada (Harcourt et al. 1955), northeastern United States (Ferro 1993), Europe (Jankowska 2006), Southeast and South Asia (Lal and Bhajan 2004; Younas et al. 2004). Yildizhan has already reported the presence of an aphrodisiac pheromone blend in male wings of *P. rapae*. The major component released from male wings was ferrulactone. Other components present in males were identified as hexahydrofarnesylacetone (HHA) and phytol (Yildizhan et al. 2009). The biological activity of the compounds in the extract was verified by coupled gas chromatography-electroantennographic detection (GC-EAD), in which they induced a signal response in antennae of conspecific female butterflies.

The main objective of the present study was to elucidate the male-specific blends released from wings of *P. rapae* butterflies of diverse geographical populations. We have compared the populations of two subspecies, namely *P. rapae crucivora* (Asian populations) and *P. rapae rapae* (European populations). Here, the pheromones from male wings were qualitatively and quantitatively analyzed and compared to the chemical profile of volatiles found in female wings. Our findings would be of particular interest for biotechnological tools in integrated pest management (IPM) strategies against this pest.

2.3 Materials and Methods

Insects - Insects were collected in fields of Taiwan (Shanua), Vietnam (Hanoi), Germany (Berlin), and the Netherlands (Wageningen). Larvae and pupae were shipped to Berlin and further reared at Humboldt-Universität, Berlin. The population of Berlin was collected in Berlin-Dahlem. Different populations of *P. rapae* were reared on Turnip cabbage (*Brassica oleracea* var. *gongylodes*, 'Delikatess Weißer', Albert Treppens & co. Samen GmbH, Germany) in gauze covered glass aquarium in an insecterium (figure 2.1), at 16h day: 8h night ratio and 26°C. In all cases, the second generation of collected insects and not the first one were used for the experiments and chemical analysis to avoid some type of bacterial and viral diseases to affect the statistics. Insects used for the pheromone extraction were 3-4 days old. Lab rearing of *P. rapae* population was on Turnip cabbage. After pupation, pupae were transferred to gauze cages in a greenhouse for eclosion of butterflies and they were fed with 10% Honey solution. Additional light was applied after sun rises for 12 hours to stimulate mating and egg laying of adults on Turnip cabbage plants. Turnip cabbage plants with eggs were returned to the insect rearing room and plants were replaced as needed.

Extraction procedure – After euthanizing the butterflies for 2h at −80°C, male and female butterflies were sexed according to their shape of abdomen and sex organs under a stereo microscope. Wings were excised from the body using surgical scissors, which were cleaned with ethanol and dried between each use. After separating the wings from the thorax, they were transferred to covered glass vials. In each vial, wings of three butterflies of one sex were transferred. Wings in the small bottles were pressed and positioned at the bottom. The first extraction was done with 2ml hexane in 4ml vials. As internal standard 10µl pentadecane (769ng/µl) was added. Samples were shaken at 500rpm for 2h at room temperature. Supernatants of hexane from all samples were removed with a Hamilton syringe and applied to Pasteur pipette columns, filled with cotton wool and sodium sulfate to remove particles and residue water. Then, the eluent was evaporated under streaming nitrogen and filled up to 300µl of volume with hexane. The extract was split and transferred to two 1.5 ml GC vials with 300µl inserts for GC-FID analysis and GC-MS analysis, respectively. A second extraction of wings was done using 2ml dichloromethane following the same steps as performed for the hexane extraction.

GC-FID and GC-MS analysis – For quantification gas chromatography coupled with flame ionization detector (A5890 series I GC system from Agilent, Germany) with auto sampler was used. Compounds were separated on a ZB-WAX plus capillary GC column (30m length, 0.25mm internal diameter, 0.25µm film thickness, Zebron, USA). The GC oven was programmed according to the following temperature program: 70°C (hold time 1min), then a ramp at 15°C/min to 220°C and a hold time of 50min. The helium flow was 1.1ml/min, the hydrogen flow was 40ml/min and the nitrogen-makeup flow was 30ml/min. A splitless injection was performed with a sample volume of 2µl. The purge flow of the FID was 11ml/min, 2min and at 50Hz. Wings compounds were quantified by using pentadecane as internal standard and calibration curves of the synthesized authentic standards, hexahydrofarnesylacetone (HHA) and phytol, obtained from Pest Control (India) Pvt Ltd (Dr. K. R. M. Bhanu). For gas chromatography coupled with mass spectrometry an Agilent (Germany) GC system of A 5890 series I equipped with an autosampler was used. Two GC capillary columns were compared, a ZB-WAX plus with a DB-1. The ZB-WAX plus column was a similar column as used for GC-FID. The DP-1 (Agilent J&W, made in the USA) was 30m length, had 0.32mm internal diameter and 0.25µm film thickness. The same temperature program as used for GC-FID was performed for comparison of results.

For quantification, the standard compounds, Pentadecane (0.83mg/µl), HHA (0.48mg/µl) and E-phytol (0.36mg/µl) were diluted to 1:1000, 1:10,000 and 1:100,000. Then the calibration curve was made, using signal intensities via GC-FID.

Statistical analysis - The statistical significance of variation in the amounts of wings pheromones among the different populations was determined using a univariate generalized linear model (GLM), followed by posthoc Fischer`s least significant difference test (LSD). All analysis was performed in SPSS (IBM SPSS Statistics 23).

Figure 2.1: Rearing of butterflies. Left side showing the glass insectarium with honey solution in Eppendorf tubes, used as food for the adult butterflies. Right side showing the glass insectarium with food plants, used for the larvae.

2.4 Results

Pheromone components in P. rapae wing extracts – In hexane and dichloromethane extracts of *P. rapae* wings similar compounds were detected (figure. 2.2 and 2.3). Ferrulactone was found to be a male-specific compound in all studied populations of *P. rapae rapae* and *P. rapae crucivora*. Also, phytol was present as a male-specific compound in *P. rapae* populations from Germany, Taiwan, and Vietnam. But phytol was below the detection level in extracts of *P. rapae* wings from the Netherlands. hexahydrofarnesylacetone (HHA) was not found to be the only male specific wing compound since it was also detected in female butterflies (figure 2.2 and 2.3), although in lower amounts.

A phytol molecular ion peak was detected at 296m/z with the abundant fragment at 71m/z and other fragments at 43m/z, 57m/z, 81m/z, 95m/z, 111m/z, and 123m/z (figure 2.4). The molecular ion peak of HHA was found at 268m/z and fragments at 58m/z, 71m/z, 109m/z, and 123m/z (figure 2.5). The presence of HHA (6, 10, 14-trimethyl-2-pentadecanone) and phytol (3, 7, 11, 15-tetramethyl-2-hexadecen-1-ol) in male wings of *P. rapae* was confirmed by GC-MS analysis.

Table 2.1: Compounds found in different *P. rapae* populations with their Retention indices after GC-MS analysis; RI1=DP-1 and RI2= ZB-wax

No.	Name of Compounds	Occurrence of compounds in male extracts of various *P. rapae* populations				RI 1	RI 2
		Netherlands	Germany	Vietnam	Taiwan		
1.	Ferrulactone	+	+	+	+	1714	2044
2.	HHA	+	+	+	+	1846	2143
3.	Phytol	-	+	+	+	2116	2625

There was no standard available for ferrulactone (cucujolide I). However, the compound identity was verified by mass spectral analysis as displayed in figure 2.6. The mass spectrum is similar to that of ferrulactone as already reported in literature (Yildizhan et al. 2009). The molecular ion peak was at 194m/z with an abundant fragment at 127m/z, related to the loss of one isoprene unit from the molecular ion. Further fragments were found at 53m/z, 61m/z, 81m/z, 99m/z, 109m/z, and 121m/z.

For further confirmation, van den Dool and Kratz retention indices (1963) for each compound was calculated as shown in table 2.1. The n-alkane series of C7-C30 (Sigma- Aldrich) and C7-C40 (Sigma- Aldrich) saturated alkanes were used. Data from two different columns were applied to calculate retention index (RI) of specific compounds. One was a non-polar column named DP-1 while the other one had 100% aqueous stability for polar compounds, called ZB-wax. Later, these Indices were compared to the Indices found in Literature for each compound with respect to the specific GC-Coloumn (Babushok et al. 2011; Mondello et al. 1995).

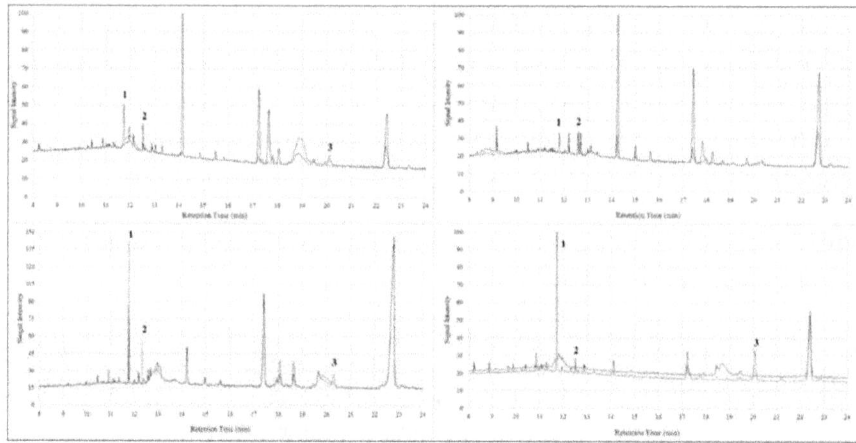

Figure 2.2: GC-FID chromatograms of hexane extracts of male and female wings of *P. rapae*. Left-up, Germany; right-up, Netherlands; Left-down, Taiwan and right-down, Vietnam. 1=Ferrulactone, 2= HHA and 3= Phytol (red=male; black=female)

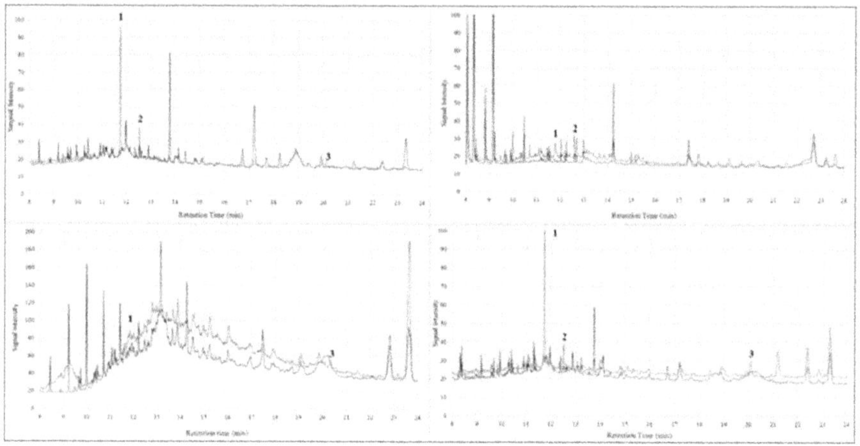

Figure 2.3: GC-FID chromatograms of dichloromethane extracts of male and female wings of *P. rapae*. Left-up, Germany; right-up, Holland; Left-down, Taiwan and right-down, Vietnam. 1=Ferrulactone, 2= HHA and 3= Phytol (red=male; black=female)

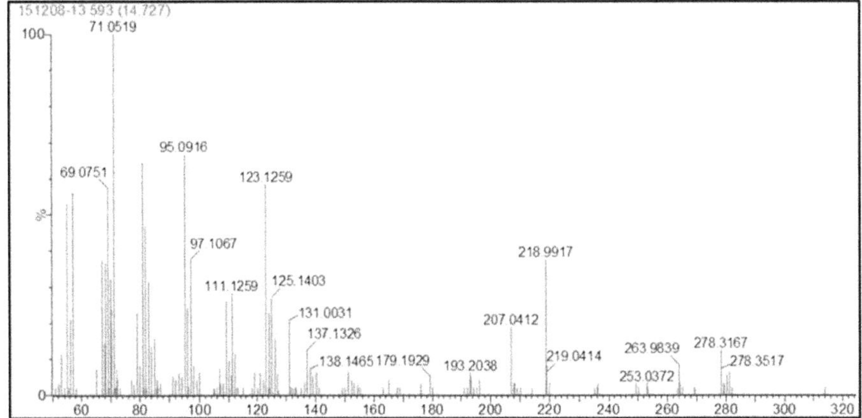

Figure 2.4: Mass spectrum of phytol

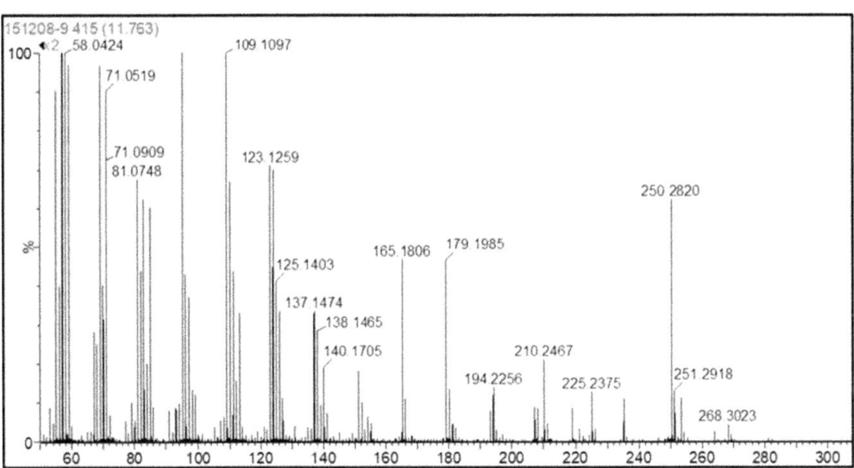

Figure 2.5: Mass spectrum of hexahydrofarnesylacetone (HHA)

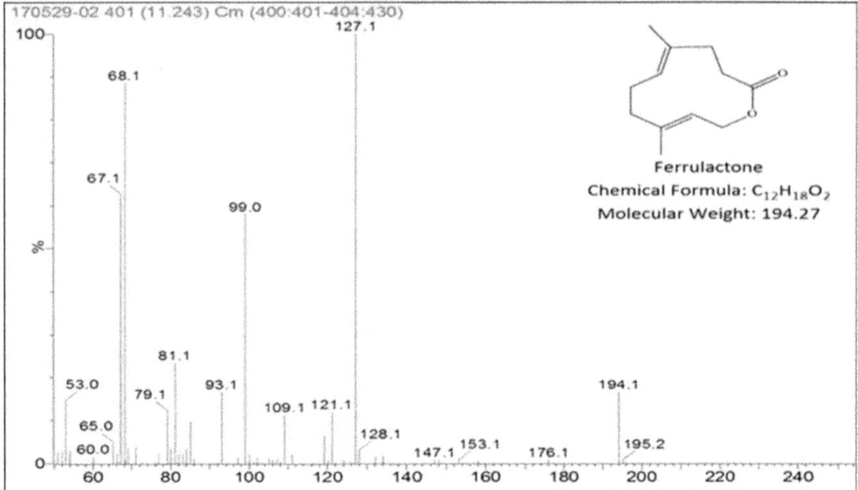

Figure 2.6: Mass spectrum of Ferrulactone

Pheromone compounds amounts - Overall, the total amount of male pheromones was higher in the Asian populations than the European ones (figure 2.7). The amount of ferrulactone, HHA and phytol found in German population was 4.0ng, 5.1ng, and 2.7ng per butterfly wings, respectively. In the Netherlands population, ferrulactone and HHA were 1.3 ng and 4.6 ng per butterfly. In the Vietnam population phytol dominated with 13.3ng, followed by HHA with 4.1ng and ferrulactone with 3.1 ng per butterfly. Also, in the same race of *P. rapae* from Taiwan, phytol was the dominating compound with 9.4ng per butterfly, followed by HHA with 5.3ng and ferrulactone with 1.2ng. Therefore, we found the double amount of ferrulactone in the German and Vietnam populations when compared to Netherland and Taiwan (figure 2.7) and phytol was present in the highest concentration in Vietnam and Taiwan. There was no significant difference in the amounts of hexahydrofarnesylacetone among all studied populations (figure 2.8). Furthermore, hexahydrofarnesylacetone was present in male and female wing extracts and could not be attributed as a male specific compound. Albeit it was found in significantly higher amounts in males in the German, Netherland, and Taiwan populations, when compared to female butterflies (figure 2.8).

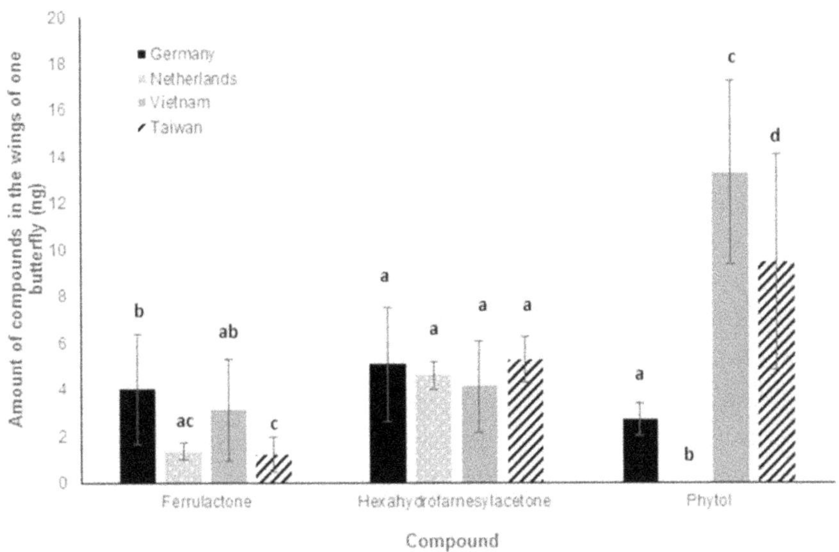

Figure 2.7: Amounts of specific compounds in the male wings extracts of *Pieris rapae* of different populations. (Different letters show significant differences within one compound among populations; Fisher's LSD P < 0.05).

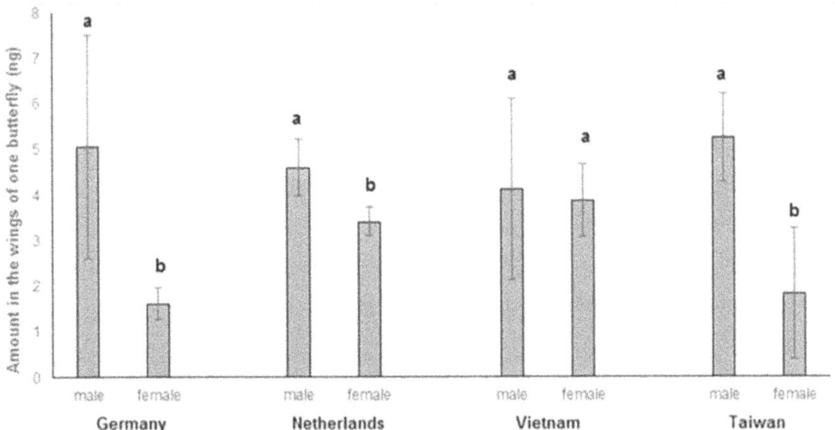

Figure 2.8: Amounts of hexahydrofarnesylacetone in the wings of *P. rapae* (male and female) of different populations. (Different letters indicate significant result among sexes; Fisher's LSD P ≤ 0.05)

2.5 Discussion

The wings pheromone profile of two Small cabbage white butterfly subspecies, that are *P. rapae crucivora* and *P. rapae rapae*, have been investigated. In this study, we have used two populations of *P. rapae rapae* from Europe (Netherland and Germany) and two populations of *P. rapae crucivora* from the Asian region (Vietnam and Taiwan). Both closely related subspecies have been studied for a long time and have shown divergence in volatile profiles of both males and females (Yildizhan et al. 2009; McQueen and Morehouse 2018). This study intended to investigate the divergence in wing volatiles profile among the four different populations of *P. rapae*, which can play an important role in understanding the reproductive isolation among different populations. According to a study, the main compounds of volatiles collected from calling males of great wax moth, *Galleria mellonella* were nonanal and undecanal and their ratio found in population from Canada were 1: 3 (Romel et al. 1992) while in the population from the USA, it was 7: 3 (Leyrer and Monroe 1973). Also, Russian populations were characterized by the two substances, but with different ratios present in the population of six different regions investigated (Lebedeva et al. 2002).

Through mitochondrial DNA sequence analysis and amplified fragment length polymorphism (AFLP) analysis, it was shown that *P. rapae rapae* is of European origin and expanded its distribution eastward to Asia (Fukano et al. 2012). The difference between the two races is that ancestral subspecies has UV-absorbing wings, and the Asian populations possess UV-reflecting wings (Obara et al. 2008). A similar study was undertaken for *Maruca vitrata* (F.) and the pheromone blend of populations from Asian and African countries was compared. Although a synthetic blend was only successfully attracting moths in Africa, but not Asian fields (Downham et al. 2004, Srinivasan et al. 2015), a geographic variation within the sex pheromone blend between Asian and West African *M. vitrata* populations could not be verified (Schläger 2015). Differently, in the present study with *P. rapae* subspecies, we observed differences within the pheromone profile ratio of European and Asian populations.

We have identified the three main wing volatile compounds: ferrulactone, hexahydrofarnesylacetone (HHA), and phytol in almost all of the populations. The male-specific pheromone of *P. rapae rapae* that is ferrulactone has been described already in the literature (Wong et al. 1983; Honda and Kawatoko 1982; Li and Mathews 2016; Yildizhan et al. 2009; McQueen and Morehouse 2018). It is also a well-known pheromone component of the stored product pest *Cryptolestes ferrugineus* Stephens (Loschiavo et al. 1986) and Phytol is diterpene alcohol, which is a widespread secondary compound of plants that is esterified to chlorophyll a

(Schulz et al. 1993). Phytol is also a male-specific compound found in detectable amounts in the wings of both subspecies (Yildizhan et al. 2009; McQueen and Morehouse 2018).

According to our study, ferrulactone was found to be in larger amounts in both populations of *P. rapae rapae*, which are from Germany and Netherlands, while phytol was present in lesser amounts in only German population. Instead, we used the same population as found in Literature (Yildizhan et al. 2009), but we could not be able to detect phytol in the population from Netherlands, probably because the amounts were below the detection levels due to different rearing conditions from different laboratories.

The main difference between the *P. rapae rapae* and *P. rapae crucivora* populations was found at the levels of phytol in male wings. While in the populations of Vietnam and Taiwan this was the male-specific dominating compound, it was not abundant or absent in the two European populations studied.

The third identified wing compound was hexahydrofarnesylacetone (HHA). It is also a part of the pheromone system of the male Danaine butterfly, *Idea leuconoe* Erichson (Schulz et al. 1993; Schulz and Nishida 1996). It was also present in the perfume blends of fifteen species of orchid bees (Zimmermann et al. 2009). Furthermore, HHA was identified in the butterfly *Bicyclus anynana* Butler, acting as a male-specific pheromone compound beside others (Nieberding et al. 2008). HHA has already been found earlier by Yildizhan (2009) in both male and female wings of *P. rapae rapae* from Netherlands. In another piece of Literature by McQueen and Morehouse (2018), it was found that HHA is a male-specific compound present in *P. rapae rapae* from Pennsylvanian as well as *P. rapae crucivora* from Japan. Our results support the earlier finding from Yildizhan; however, females of all four populations in this study contained HHA as well, albeit in lower amounts than males. Therefore, we could not confirm that HHA is a male-specific compound. Furthermore, our statistical analysis showed that HHA is found in all of our four tested populations in equal amounts, with a nonsignificant difference. Here, our results are not in accordance with McQueen and Morehouse (2018) who found HHA in higher amounts in the population collected from Pennsylvania when compared to the Japanese population.

The quantification of pheromone components is a matter of considerable impact because it is very advantageous for bioassays and mass trapping field trails. Lu et al. (2013) recently found pheromone blend variation between two Chinese *M. vitrata* populations in field studies. A pheromone ratio of 100:10:80 (EE10,12-16:Ald: EE10,12-16:OH: E10-16: Ald) attracted the maximum number of males in Wuhan (total number of males caught per trap: 19.5±3.6 SD), whereas 100:10:10- blend was the most attractive in Huazhou (17.8±1.9SD). Besides *P. rapae*, quantitative pheromone analysis and bioassays are available for many lepidopteran species, such as *Dolbina tancrei* Staudinger (Uehara et al. 2013). The sex pheromone of this species consists of (9E,11Z)-9,11-pentadecadienal and (9Z,11Z)-9,11-pentadecadienal, with the highest male catches observed for a ratio of 90:10 blend. According to our quantitative analysis, the total amount of pheromone compounds produced by *P. rapae rapae* populations is lower than populations of *P. rapae crucivora* under similar rearing conditions. Our findings indicate geographic variation in the sex pheromone blend between Asian and European *Pieris* populations.

According to a bioassay carried by Yildizhan, it was proved that additional spray of these three compounds on the wings of males increases the percentage of mating (Yildizhan et al. 2009). However, it could be concluded that these three compounds have aphrodisiac properties and can be used as pheromones lures in field trials. Both Asian populations showed a similar pheromone profile, therefore the same pattern of these compounds could be tried as pheromone traps for populations of Asian origin. Since both populations of European subspecies have a different profile, different lures for trapping need to be designed for integrated pest management.

3 Electroantennographic responses of female *P. rapae* subspecies to the male wings' volatiles

3.1 Abstract

Since the interest in a chemical-free environment is rising, the integrated pest management strategies like, mating disruption has become ubiquitous in the field of agriculture. Electroantennography is a very well-known and desirable technique used to find out the responses of pests to synthetic chemical compounds, which can easily apply to the mating disruption process in agricultural fields.

In the present study, we checked out the responses of antennae of *P. rapae crucivora* (Taiwan) females to single synthetic compounds, hexahydrofarnesylacetone (HHA) and E-phytol, found in male wings extracts via manual puffing. We found out that the female showed clear responses to pure synthetic HHA as well as to 2:1 (HHA: E-phytol) ratio, but not E-phytol alone. The GC-EAD analysis was carried out to observe the responses of females to ferrulactone (male-specific), HHA, and E-phytol. *P rapae rapae* (Germany), *P. rapae crucivora* (Taiwan), and *P. rapae crucivora* (Vietnam) were used for this analysis. All three populations of *P. rapae* females respond to ferrulactone as well as HHF. The population from Germany and Vietnam also showed response to E-phytol. Summarizing the results, ferrulactone and HHA could be used as pheromone lures for both the subspecies to carried out field bioassays to test the mating disruption strategy.

Furthermore, we have tested the antennal responses of *P. rapae rapae* (Germany) female to the *P. rapae crucivora* (Taiwan) male wings extracts and vice versa, to check the distinguishing ability of females to find their conspecifics. As a result, *P. rapae crucivora* (Taiwan) did not show and respond to the allospecific male wings extract but *P. rapae rapae* (Germany) females respond to allospecific as well as conspecific wings extracts.

3.2 Introduction

The electrical potential between the two ends of insect antennae could be captured as a response to a specific odor and was first named as electroantennogram or EAG in the middle era of the twentieth century (Schneider 1957). Over the years, the application of EAG has been widely used to identify plant volatiles, insect pheromones, and foraging behavior of butterflies (Cosse et al. 1995; Honda et al. 1999; Conner et al. 1980).

The olfactory preferences of some tropical butterflies have been studied to observe their foraging behavior. As a result of GC-EAD analysis, fifteen aliphatic esters were shown to be detected by the butterflies' sensory apparatus such as proboscis and antennae (Sourakov et al. 2012). Foraging behavior of *P. rapae* was also studied and it was found out that 2 phenylethanol and phenylacetaldehyde stimulate their foraging behavior (Honda et al. 1998; Ômura et al. 1999). The behavioral responses of foraging adults of *Kallima inachus* (Boisduval) to five volatile components found in the six host fruits was also done by using the EAG technique, to determine the cues used by foraging adults (Tang et al. 2013). The olfactory responses of Indian meal moth, *Plodia interpunctella* to fractions of wheat extracts and a single pure compound, nonanal was checked. The fractions containing 27 compounds as well as a single compound, nonanal, both were found to be highly attractive to *Plodia interpunctella* (Uechi et al. 2007). Electroantennographic responses of adult butterflies of six nymphalid species and the results suggested that in food selection, butterflies use a specific olfactory system to perceive the major food-derived volatiles. All the six species showed high response signals to the several flower scent aromatic compounds, while three of them also showed responses to the fermented products of those compounds (Omura and Honda 2009).

EAG response profiles of five different insect species showed different species-specific antennal responses to 20 volatile compounds tested and most of the compounds could be distinguished by comparing the response spectra (Park et al. 2002). Two electrophysiologically active pheromone compounds, tetradecyl acetate (14: OAc) and Z9-tetradecenyl acetate (Z9-14: OAc) were identified from pheromone gland extracts of female cossid moth, *Coryphodema tristis*, and an additional compound (Z9-14: OH) from headspace samples. By using a specific ratio of these compounds in traps in field trails, a significant number of males were caught (Bouwer et al. 2015).

Identification of female sex pheromone of carpenter moth, *Cossus insularis* by GC-EAD analysis had played an important role in the development of a commercial mating disruption product. Two compounds (E)-3-tetradecenyl acetate and (Z)-3-tetradecenyl released by live females elicited EAG responses from the antennae of male moths. From the identification of female sex pheromones to the acceptance of a synthetic version of that sex pheromone as an agrochemical, EAG studies played a significant role (Chen et al. 2006; Hoshi et al. 2016).

In the present study, we wanted to check the responses of antennae of *P. rapae rapae* and *P. rapae crucivora* females to single synthetic compounds found in male wings extracts via either manual puffing EAG or coupled GC-EAD techniques. These compounds could be used as pheromone lures to perform the field bioassays which in turn, could help to control this pest insect via mating disruption strategy in agricultural fields. Furthermore, we tested two different subspecies, *P. rapae*

rapae, and *P. rapae crucivora* females, to check whether they could distinguish between the allospecific and conspecific males.

3.3 Materials and Methods

Insects - Insects were collected in fields of Taiwan (Shanua), Vietnam (Hanoi), and Germany (Berlin). Larvae and pupae were shipped to Berlin and further reared at Humboldt-Universität, Berlin. The population of Berlin was collected in Berlin-Dahlem. Different populations of *P. rapae* were reared on Turnip cabbage (*Brassica oleracea* var. *gongylodes*, 'Delikatess Weißer', Albert Treppens & co. Samen GmbH, Germany) in gauze covered glass insecterium, at 16h day: 8h night ratio and 26°C in insects rearing room. In all cases, the second generation of collected insects and not the first one were used for the experiments and chemical analysis to avoid some type of bacterial and viral diseases to affect the statistics. Insects used for the pheromone extraction were 3-4 days old. Females used for EAG analysis were 6-8 days old. Lab rearing of *P. rapae* population was on Turnip cabbage. After pupation, pupae were transferred to gauze cages in a greenhouse for eclosion of butterflies and they were fed with 10% Honey solution. Additional light was applied after sun rises for 12 hours to stimulate mating and egg laying of adults on Turnip cabbage plants. Turnip cabbage plants with eggs were returned to the insect rearing room and plants were replaced as needed.

Extraction procedure – After euthanizing the butterflies for 2h at −80°C, male and female butterflies were sexed according to their shape of abdomen and sex organs under a stereo microscope. Wings were excised from the body using surgical scissors, which were cleaned with ethanol and dried between each use. After separating the wings from the thorax, they were transferred to covered glass vials. In each vial, wings of three butterflies of one sex were transferred. Wings in the small bottles were pressed and positioned at the bottom. The first extraction was done with 2ml hexane in 4ml vials. As internal standard 10µl pentadecane (769ng/µl) was added. Samples were shaken at 500rpm for 2h at room temperature. Supernatants of hexane from all samples were removed with a Hamilton syringe and applied to Pasteur pipette columns, filled with cotton wool and sodium sulfate to remove particles and residue water. Then, the eluent was evaporated under streaming nitrogen and filled up to 300µl of volume with hexane. The extract was split and transferred to two 1.5 ml GC vials with 300µl inserts for GC-FID analysis and GC-EAG analysis respectively. A second extraction of wings was done using 2ml dichloromethane following the same steps as performed for the hexane extraction.

Manual Electroantennographic analysis - EAG responses were recorded from the antennae of 4-8 days old adults of both sexes. The antennae were freshly sedated by CO2. The antennae were moved carefully into the holder by touching the cut end of the antenna with a needle wetted with insect Ringer solution. The ringer solution was used as previously described by Schott et al. It consists of 9 g/l NaCl, 0.2 g/l KCl, 0.25 g/l MgCl2 and 1 g/l glucose. The antenna was slowly inserted into the slit between the two reservoirs filled with ringer solution. These reservoirs contain silver electrode with a Pt wire. The EAG amplifier, filter, A/D converter (IDAC 2, Syntech, Kirchzarten, Germany), charcoal filter, humidifier, and a valve for controlling the monitoring air puffs were mounted in a portable rack. For air puffs, syringes were prepared to contain a piece of filter paper with 1 cm^2 area, treated with 5 µl of test solvent.

GC-FID analysis – For identification of compounds in the wings extracts, gas chromatography coupled with flame ionization detector (A5890 series I GC system from Agilent, Germany) with auto sampler was used. Compounds were separated on a ZB-WAX plus capillary GC column (30m length, 0.25mm internal diameter, 0.25µm film thickness, Zebron, USA). The GC oven was programmed according to the following temperature program: 70°C (hold time 1min), then a ramp at 15°C/min to 220°C and a hold time of 50min. The helium flow was 1.1ml/min, the hydrogen flow was 40ml/min and the nitrogen-makeup flow was 30ml/min. A splitless injection was performed with a sample volume of 2µl. The purge flow of the FID was 11ml/min, 2min and at 50Hz. Wings compounds were quantified by using pentadecane as internal standard and calibration curves of the synthesized authentic standards, hexahydrofarnesylacetone (HHA) and E-phytol, obtained from Pest Control (India) Pvt Ltd (Dr. K. R. M. Bhanu).

GC-EAD analysis- For gas chromatography coupled with electroantennographic detector (A5890 series I GC system from Agilent, Germany) with auto sampler was used. Compounds were separated on a ZB-WAX plus capillary GC column (30 m length, 0.25 mm internal diameter, 0.25 µm film thickness, Zebron, USA). The GC oven was programmed according to the following temperature program: 70°C (hold time 1min), then a ramp at 15°C/min to 220°C and a hold time of 50 min. The helium flow was 1.1 ml/min, the hydrogen flow was 40 ml/min and the nitrogen-makeup flow was 30 ml/min. A splitless injection was performed with a sample volume of 2 µl. Dilutions of synthesized authentic standards, hexahydrofarnesylacetone (HHA) and E-phytol were prepared and 0.1µg/ml of both the compounds was used for puffing EAG analysis.

GC-separated volatiles were then forced to a mixing tube, where it mixed with cleaned air with 60-80% humidity. This mixing tube has a fixed temperature of 250°C. the volatiles were then transferred to the antennae holder. Freshly cut antennae from CO_2-sedated females were inserted into the slit between two reservoirs of holder (figure 3.1), filled with Ringer solution (9 g/l NaCl, 0.2 g/l KCl, 0.25 g/l $MgCl_2$ and 1 g/l glucose). Insect-antenna holder was designed for this purpose with 7mm, 8mm and 9 mm crevices (figure 3.1) to hold the about 1cm long antenna among the ringer reservoirs with centered Ag/AgCl-sensors. These reservoirs contain silver electrode with a Pt wire. The potential difference between the electrodes produces a signal which is detected by the electroantennographic detector. Antennal responses were amplified 10× by using a Universal AC/DC Probe connected to an IDAC 4 signal converter (Syntech).

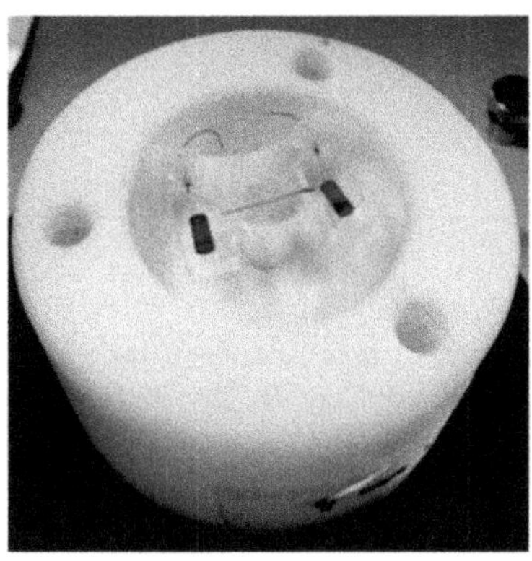

Figure 3.1: A *P. rapae* female antennae placed between the two electrodes of antennae holder for electroantennographic analysis

3.4 Results

EAG assay to check the pure synthetic compounds and their ratio - We conducted EAG studies to test the bioactive effects of single synthetic compounds, HHA and E-Phytol. The *P. rapae rapae* from Taiwan have been used for this experiment. We used allyl isothiocyanate – a characteristic host-plant volatile as a positive control. First the response of the positive control, that is isothiocyanate was checked in paraffin oil as well as in hexane to investigate whether the whole process of EAG puffing worked (figure 3.2). The induced signals in paraffin oil (about 0.15mV) were higher than in hexane (about 0.05mV).

Then the antennae of female butterflies were used to check their response to the male butterflies' wings extracts and other two compounds, that are HHA and E-Phytol and also a mixture of these two compounds with 2:1 (HHA: E-Phytol) ratio was checked. We got the highest EAG response as a result of a single puff of male wings extract. It was about 0.2mV. EAG responses elicited by puffing HHA and also 2:1 ratio of HHA and E-phytol showed smaller signals than the male wings extract. While with E-phytol alone, we did not get any EAG signal (figure 3.3).

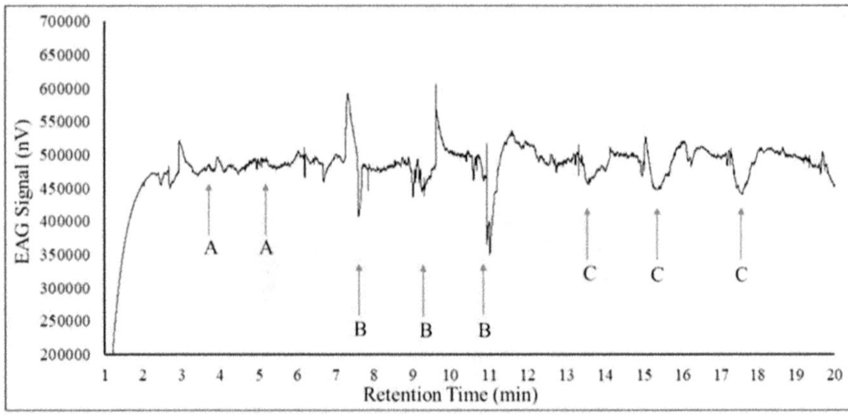

Figure 3.2: EAG response of female *P. rapae* antennae to the positive control allyl isothiocyanate (Taiwan female, 7 day old) A-paraffin oil; B- allyl isothiocyanate in paraffin oil; C- allyl isothiocyanate in hexane.

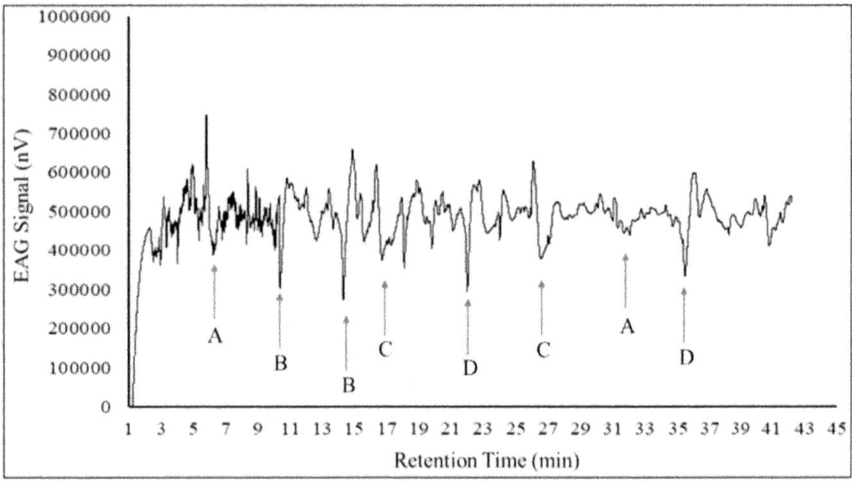

Figure 3.3: EAG response of female *P. rapae crucivora* antennae (Taiwan, 6 day old) to A-air; B- male wings extract; C- HHA and E-phytol mixture (2:1); D-(0.1µg/ml)100% HHA.

EAG assay to check female response to allospecific and conspecific male - We conducted EAG assay in which antenna of female from both the subspecies were used to test the wings extracts of *P. rapae rapae* males as well as *P. rapae crucivora* males. EAG signals with *P. rapae crucivora* female antenna elicited by puffing the wings extract of male *P. rapae crucivora* were higher than the signals appeared as a result of wings extract of male *P. rapae rapae*. The response of *P. rapae crucivora* female with wings extract of male *P. rapae rapae* were like the response with air puffed with the syringe as negative control (figure 3.4). We got the EAG response signals with the female antennae of *P. rapae rapae* for both the extracts. But the EAG signals produced by puffing the wings extract of male *P. rapae rapae* were higher than the signals appeared as a result of wings extract of male *P. rapae crucivora* (figure 3.5). The responses captured by the antenna of *P. rapae rapae* females always showed a lot of noise.

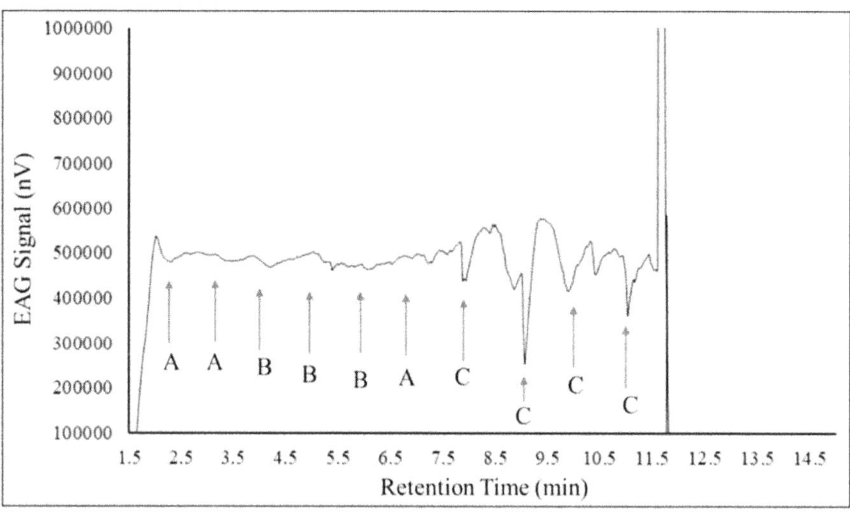

Figure 3.4: EAG response of 6 days old female *P. rapae rapae* antennae (Taiwan) to the wings' extracts of allospecific and conspecific males. (A-Air; B- male *P. rapae crucivora* (German) wings extract; C- male *P. rapae rapae* (Taiwan) wings extract)

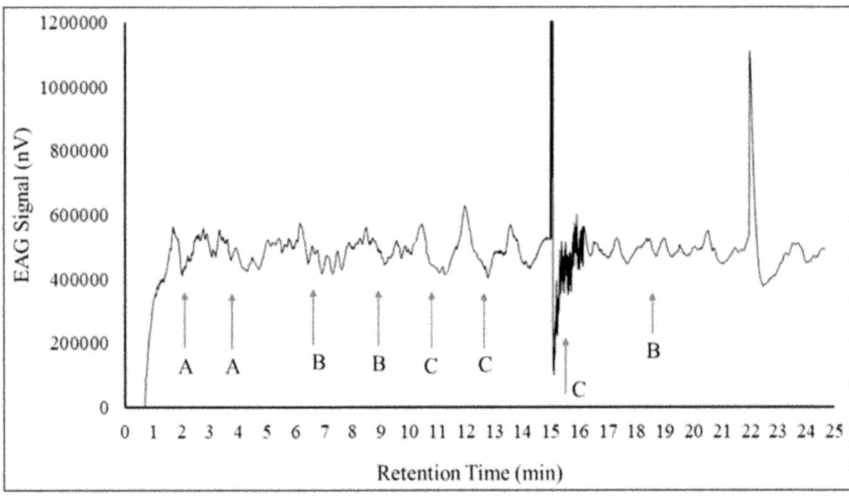

Figure 3.5: EAG response of 6 days old female *P. rapae crucivora* antennae (German) to the wings' extracts of allospecific and conspecific males. (A-Air; B- male *P. rapae rapae* (Taiwan) wings extract; C- male *P. rapae crucivora* (German) wings extract)

GC-EAG studies - We conducted also successful GC-EAG studies with *P. rapae crucivora* (Taiwan), *P. rapae crucivora* (Vietnam) and *P. rapae rapae* (Germany), but we found a lot of noise signals for most of the samples. For this study, we used female antenna to check its response towards the compounds found in the wings extracts of their conspecific males. The GC-EAD chromatogram were then compared with the GC-FID chromatograms as described in chapter 3. In GC-EAD chromatogram of *P rapae crucivora* (Taiwan), we found a clear response to ferrulactone at a retention time of 11.8 min, that is a male specific pheromone. We also got a small EAG signal for HHA which could be seen at 12.67 min retention time. But we did not find any response to E-phytol at 19.5 min (figure 3.6). GC-EAG chromatogram of *P rapae crucivora* (Vietnam) female also showed two signals with highest amplitude, that are, for Ferrulactone and HHA, as Taiwan population. We got many other signals too, even here we can see a signal for E-phytol between 18.7 to 19. 5 min (figure 3.7).

All the GC-EAD chromatograms of *P. rapae rapae* (Germany) female showed a lot of noise as it was with puffing. Here we could see clear response of female to the three compounds of wings extract, that are ferrulactone, HHA and also E-phytol (figure 3.8). Here we could recognize a better signal for E-phytol as in Vietnam population. Furthermore, in GC-EAG studies, the female antenna responded to more than the three known pheromone compounds.

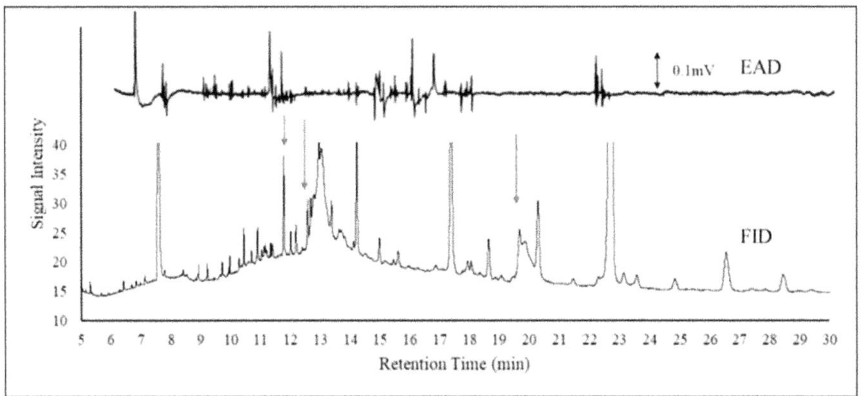

Figure 3.6: GC-EAD showing the response of female antennae of *P. rapae crucivora* (Taiwan) towards the GC-FID of male wings extract of same population.

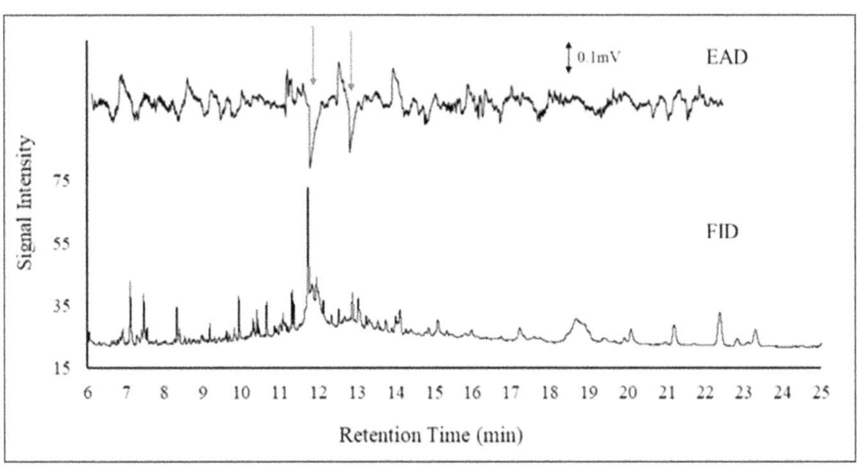

Figure 3.7: GC-EAD showing the response of female antennae of *P. rapae crucivora* (Vietnam) towards the GC-FID of male wings extract of same population.

Unfortunately, we were not able to proceed to the statistical analysis due to the smaller number of good EAG spectra. It was really very difficult to develop EAG studies for cabbage white butterflies as they are very sensitive insects and EAG technique is itself very sensitive and many factors can affect the results, such as temperature and humidity of room.

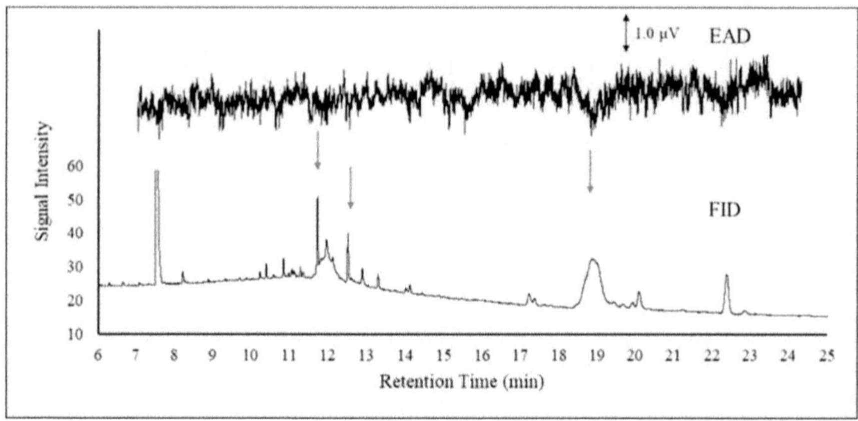

Figure 3.8: GC-EAD showing the response of female antennae of *P. rapae rapae* (Germany) towards the GC-FID of male wings extract of same population.

3.5 Discussion

In the present work, we have studied the electroantennographic responses of female butterflies towards some specific compounds that were found in the wing extracts of male butterflies in specific proportions. These female responses against specific chemicals could be beneficently used in the field trial experiments and agricultural practices, which could lead to control pest insects by mating disruption mechanism. Mating disruption has been practiced for a long time against many insect pests to reduce biotic stress on crops (Cardé and Minks 1995; Hegazi et al. 2010; Matsumoto et al. 2007; Witzgall et al. 2010; Aono et al. 1989; Yaginuma 1984; Escofet et al. 2005; Natale and Pasqualini 1999).

The two compounds HHA and E-phytol were previously identified and quantified from the male and female wings of *P. rapae crucivora* (Taiwan and Vietnam) (Aftab et al. 2016). To test the activity of synthetic HHA and E-phytol via puffing, we used *P. rapae crucivora* (Taiwan). We observed that the female antenna responds to the pure synthetic HHA and 2:1 ratio of HHA and E-phytol, but no response to pure E-phytol. Then we carried out GC EAG analysis to check the response of single compounds found in the wings. For that purpose, we used two populations of *P. rapae crucivora* (Taiwan and Vietnam) and also *P. rapae rapae* (Germany). The EAG responses of females of all three populations showed very clear attraction to Ferrulactone, which is a male-specific compound, and also to HHA. The two populations *P. rapae crucivora* (Vietnam) and *P. rapae rapae* (Germany) show response towards E-phytol. The same result of EAG analysis of *P. rapae rapae* (Netherlands) has been reported in the literature (Yildizhan et al. 2009). But *P.*

rapae crucivora (Taiwan) showed no response to E-phytol in GC-EAG as well as during puffing. Thus, we can conclude here that both populations of *P. rapae crucivora* were found not to be identical. Hence in this chapter, we found out that the concentration of some specific wing volatiles in both of these populations is different. Concerning the female antenna responses, we can say that *P. rapae crucivora* (Vietnam) is somehow identical to *P. rapae rapae* (Germany).

We carried out another small experiment to check the response of female antenna to the male wings extracts of the same subspecies as well as to the other. Both the subspecies, *P. rapae crucivora* (Taiwan) and *P. rapae rapae* (Germany), have been investigated. The olfactory (EAG) reaction of *P. rapae crucivora* (Taiwan) female showed that they identified their conspecifics male extract well and they did not respond to the allospecific male extract, whereas the *P. rapae rapae* (Germany) females respond to both the allospecific as well as conspecific male's extract. According to a mate recognition research, *P. rapae* and *P. melete* males approach females and attempted to copulate indiscriminately, but the females of both species could discriminate very efficiently and discontinued the mate-refusal posture only for their conspecific males (Ohguchi and Hidaka 1988). Summarizing our results, the antenna of female *P. rapae crucivora* were found to be more specific to find their conspecifics than the *P. rapae rapae*. Although the behavioral role of the single compound was not evaluated in this previous study, the attraction of gravid females to mixtures containing this compound was enhanced, suggesting a synergistic impact on behavior (Uechi et al. 2007).

4. Quantification of volatile compounds in the wings of *Pieris brassicae* and attraction tests of compounds in behavioral assay

4.1 Abstract

The large cabbage white butterfly *P. brassicae* is a worldwide pest. It uses the wings volatiles to attract their conspecifics before courtship. These volatiles were extracted from the wings and GC-MS analysis was carried out to identify and quantify the compounds. brassicalactone was found in only male wings extracts (25.45 ng/µl ± 25.90 in one butterfly). hexahydrofarnesylacetone (HHA), isophytol, and E-phytol were found in both male and female butterflies. Among all these four compounds, E-phytol was the major constituent of the scent bouquet of both males and females. E-phytol was produced ten-fold more in males than females. An electroantennographic response of female antennae was observed against these compounds. The compounds were thought to play a role in courtship. Therefore, we used different concentrations of E-phytol along with HHA and isophytol in a behavioral bioassay. Single compounds did not induce any significant response. According to the Wilcoxson test, with one specific ratio (1:1:70; HHA: isophytol: E-phytol) of these three compounds, we have observed the significant response of female butterflies same as the response with the male wings extracts and live male butterflies. Hence, the ratio (1:1:70; HHA: isophytol: E-phytol) would be suggested to use in the field trial experiments of mating disruption of *P. brassicae*. From this study, we can conclude that Large cabbage whites use a specific ratio of specific compounds to attract their conspecifics before courtship.

4.2 Introduction

Male Pierids produce specific compounds in the scales, arranged in the anteroposterior direction of the wings (Yoshida et al. 2000). These compounds induce a behavioral response in females that helps them to communicate with males, in turn promoting courtship. These wings compounds are known as aphrodisiac pheromones and have been studied in detail in various species of the Pieridae family. In 1973, it was documented that males of *Coilas eurytheme* Boisduval and *Colias philodice* Godard emit a scent used by the females for species discrimination (Taylor 1973). Later, the blend compounds of *C. philodice* were identified that were n-hexyl myristate, n-hexyl palmitate, n-hexyl stearate, n-heptacosane, and n-nonacosane combined in 7:45:9:22:16 ratio. The *C. eurytheme* blend was made by mixing n-heptacosane, 13-methyl heptacosane, and n-nonacosane in 9:80:11 ratio (Grula et al. 1980). Many assays were established to demonstrate the existence of aphrodisiac pheromones in various Pierid species (Rutowski 1977a; 1977b; 1980; Yildizhan et al. 2009). In 2009, wing compounds of *P. rapae* and *P. brassicae* have been identified. The males of *P. rapae* produce ferrulactone while *P. brassicae* males produce brassicalactone as a male-specific aphrodisiac pheromone (Yildizhan et al. 2009).

The mode of action of specific compounds could be studied and proved through electroantennographic signals of antennae of conspecifics of insects. Different blends of compounds in various proportions have been used for many moth species to check the response of their antennae. The EAG response of *Choristoneura rosaceana* Harris (Lepidoptera: Tortricidae) male antennae to the pheromone gland compounds was studied and compared among several regions such as British Columbia, Michigan, Ontario, New York, and Quebec. A field trapping experiment was conducted in each of these locations to determine the effect of four pheromone-gland compounds on the numbers of moths captured. The relative amount of (Z)-11-tetradecenyl acetate, one of the pheromone compounds was greatest in moths from New York and smallest in moths from Ontario, whereas the relative amount of another compound, (E)-11-tetradecenyl acetate was greatest in moths from Ontario and smallest in moths from British Columbia (El-Syed et al. 2003). Another research on a certain ratio of pheromone glands compounds of female *Maruca vitrata* (F.) leads to successful field trapping experiments in Benin (Downham et al. 2004). However, this ratio of blend did not attract males in Taiwan (Schläger et al. 2012), Thailand, and Vietnam (Srinivasan et al. 2015). Therefore, there is a need for quantitative investigation on pheromone banquet produced by Lepidoptera species like *P. brassicae*. It is considered a severe pest of *Brassica* plants worldwide. It was reported to be one of the most destructive and cosmopolitan pests of cruciferous crops in India, where it causes 40% of the damage per year (Hasan and Ansari 2010; 2011). The larvae can completely defoliate and kill a plant (Hill 1987). In Turkey, it caused damages to cabbage crops about 40% and to the cauliflower fields about 27% (Atalay and Hincal 1992).

In this study, we have quantified some specific aphrodisiac compounds of *P. brassicae* and the response of male pheromone lures by females in a bioassay was also determined. Our findings will furthermore be used to develop a pheromone lure that can be used for mass trapping strategy to control *P. brassicae* adults in agricultural fields.

4.3 Materials and Methods

Insects - Larvae and eggs were collected from the trap fields of cabbage plants at JKI, Berlin. They were then brought to HU, Berlin and reared in insectarium (16h photoperiod) at 26°C. The glass aquariums were used to place larvae until they become pupae. They were fed with Turnip cabbage (*Brassica oleracea* var. *gongylodes*) plants from Albert Treppens & Co. Samen GmbH, Germany.

Extraction of Aphrodisiac pheromones - Sex separation was done according to Feltwell with the help of a microscope. Butterflies were selected 4 - 5 days after eclosion. Wings were dissected from male and female butterflies and transferred to covered vails. In one vail, three butterflies' wings were transferred. The wings in the small vails were a little bit pressed so they will be at the bottom and not on the walls side. The first extraction was done by 2 ml Hexane in 4 ml Vials, so on the wings, 2 ml Hexane and 10 µl Pentadecane (as internal standard) were added. All samples were shaken at 500rpm for 2 h at room temperature. Supernatants of hexane from all samples were separated and filtered in the Pasteur pipettes columns filled with cotton and sodium sulfate (Sigma-Aldrich). The eluent was then evaporated to get about 300 µl volume with the help of nitrogen flow. Then the extract was transferred to two 1.5 ml GC-vails (150 µl to each vail with 100 µl inserts). One of the vails was used for GC-FID analysis while the other vail was used for GC-MS analysis. The second extraction was done with 2 ml dichloromethane with the same procedure as described for the hexane extraction.

Gas chromatography coupled with Mass chromatography (GC-MS-TOF) - A 5890 series I GC system (Agilent technologies, Germany) equipped with an autosampler (Gerstel, Germany) was used for the analysis. The column, named HP-5ms (30 m length, 0.250 mm internal diameter, 0.25 µm film thickness) from Agilent J&W, USA was used. GC oven was programmed for the following temperature program: 70°C (hold time 1 min), then increasing temperature with 15°C/min to 220°C (hold time, 50 min). A helium flow of 1.1 ml/min, hydrogen flow of 40 ml/min and nitrogen-makeup flow of 30 ml/min was used. Autosampler injected the volume of 1 µl spitless.

Electroantennographic analysis - EAG responses were recorded from the antennae of 4-8 days old adults of both sexes. The antennae were freshly sedated by CO_2. The antennae were moved carefully into the holder by touching the cut end of the antenna with a needle wetted with insect Ringer solution. The ringer solution was used as previously described by Schott et al (2013). It consists of 9 g/l NaCl, 0.2 g/l KCl, 0.25 g/l $MgCl_2$ and 1 g/l glucose. The antenna was slowly inserted into the slit between the two reservoirs filled with ringer solution. These reservoirs contain silver electrode with a Pt wire. The EAG amplifier, filter, A/D converter (IDAC 2, Syntech, Kirchzarten, Germany), charcoal filter, humidifier, and a valve for controlling the monitoring air puffs were mounted in a portable rack. For air puffs, syringes were prepared to contain a piece of filter paper with 1 cm^2 area, treated with 5 µl of test solvent.

Behavioral assay - Wind tunnel assay was conducted by using both live female butterflies and their wings extracts as pheromone source. The custom-made wind tunnel (80 cm long × 35.5 cm high × 37.5 cm wide) (figure 4.1) was made of glass, except for one long side which was made of polypropylene with two closable windows to be able to exchange the butterflies and compounds. An airflow of 24-26 cm/s was generated by an axial fan (REW 150/2, axial in-duct fan 150 mm, 6 in 1 PH, Helios, Villingen-Schwenningen, Germany) in the wind tunnel. Incoming air was cleaned using active charcoal (double filter for range hoods, Menz & Könecke GmbH, Elka Pieterman Group, Niederkrüchten, Germany).

Every experimental day before starting, the wind tunnel was cleaned with 70% ethanol (Carl Roth GmbH+ CO. KG, Karlsruhe, Germany) while air was flowing through the tunnel. During the experiment, 2 µl of pheromone compounds solutions were absorbed on an artificial butterfly made of filter paper (Whatman 589/3,110mm) was used as a source. The ratios 1:1:10, 1:1:20, 1:1:30 and 1:1:70 of HHA, isophytol and phytol were prepared. All the ratios were prepared by using 0.2 µg of compounds in hexane. After pipetting the solutions on filter paper, it was used for the bioassay after 3-5 sec, so that solvent was evaporated. Then the butterflies were released with the help of a 250 ml cup, at a point 40 cm away from the solvent source.

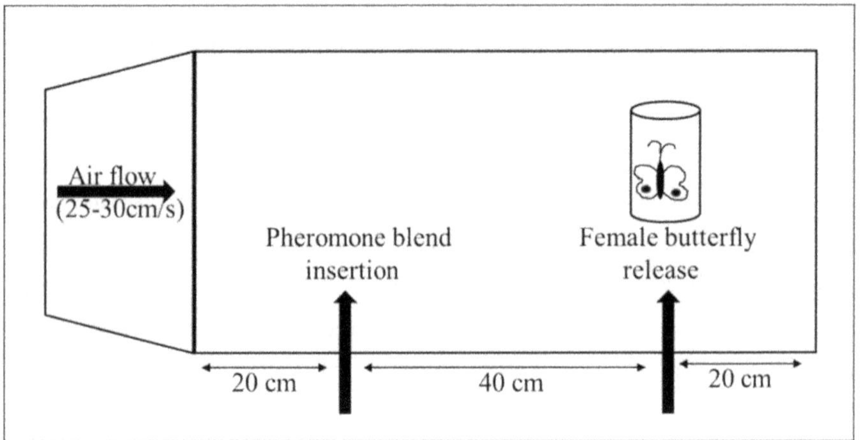

Figure 4.1: Schematic diagram of Wind tunnel used in behavioral assay.

Statistical analysis of data - The statistical significance of variation in the amounts of aphrodisiac compounds among different populations was determined by using univariate generalized linear model (GLM), followed by posthoc test and Fishers least significant difference (LSD). For the bioassay results, we have used the non-parametric Wilcoxon test to calculate the significance. All analysis was performed in SPSS (IBM SPSS Statistics 23).

4.4 Results

Chemical analysis of wings pheromone extracts - Wings of the German population of *P. brassicae* were extracted with two solvents, hexane, and dichloromethane. GC chromatograms of hexane-extracted male and female wings are shown in figure 4.2. We found four compounds. E-Phytol (A) at 14.1 min, hexahydrofarnesylacetone (B) at 11.4 min and isophytol (C) at 12.2 min were found in males as well as in females extracts, showed in table 4.1. The first three compounds A, B and C were found in lower amounts in females when compared to males' extracts (figure 4.2). Only brassicalactone (D) was found at the retention time 12.7 min and recognized as a male-specific compound.

Confirmation of aphrodisiac compounds in Pieris brassicae by GC-MS-TOF - All of the three compounds A, B, and C were confirmed by GC-MS-TOF spectral analysis (figure 4.2). Their mass spectra and fragmentation pattern were compared with their standards. For confirmation, Kovats retention indices were also calculated (table 4.1). N-alkane series of C7-C30 and C7-C40 (Sigma-Aldrich) saturated alkanes were used for the calculation of retention indices of compounds (Kovats 1958).

The mass spectral analysis of brassicalactone (figure 4.3) showed same fragmentation pattern as reported by Yildizhan (2009). It showed a peak at 68 m/z which is a characteristic peak of Isoprene unit in terpenes. The molecular ion peak was at 262 m/z, which is the exact molecular mass of this compound. Two more peaks at 194 m/z and 127m/z proved the characterization of removal of one isoprene unit from the molecular ion and further. The mass spectrum of isophytol was shown if figure 4.4. Its molecular ion peak was found at 278m/z. The mass spectra of HHA, and E-phytol were discussed in chapter 2. The peaks in the mass spectra were compared in the spectra found in the library of software of GC-MS-TOF and were confirmed by comparing them with standard.

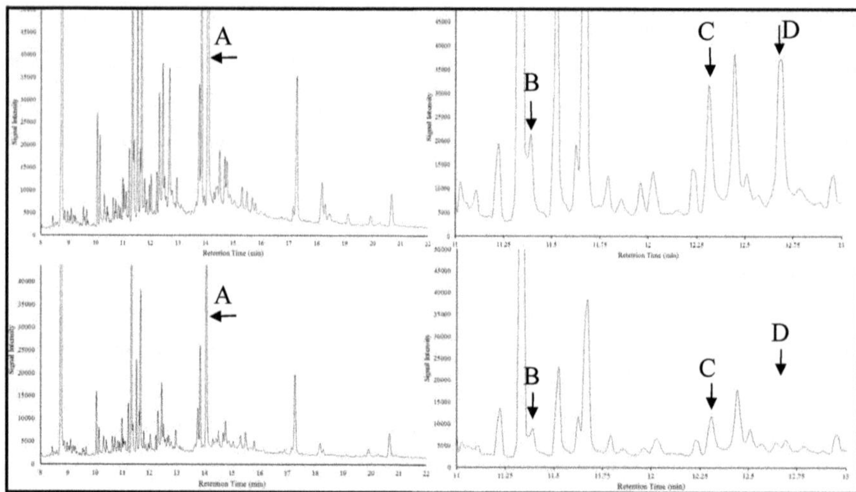

Figure 4.2: GC chromatograms of wings extracts of male and female wings of *P. brassicae*. (Upper male; lower: female; A: E-phytol: B: Hexahydrofarnesylacetone; C: Isophytol; D: Brassicalactone)

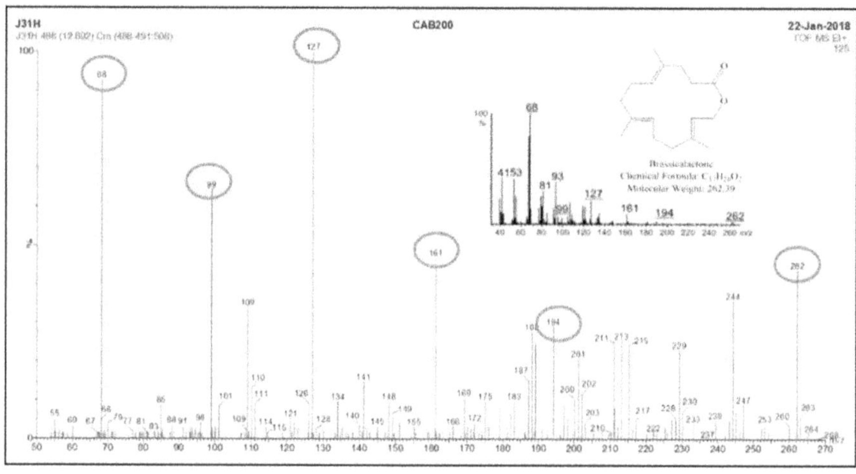

Figure 4.3: Mass spectra of brassicalactone showing 262 m/z as molecular ion peak

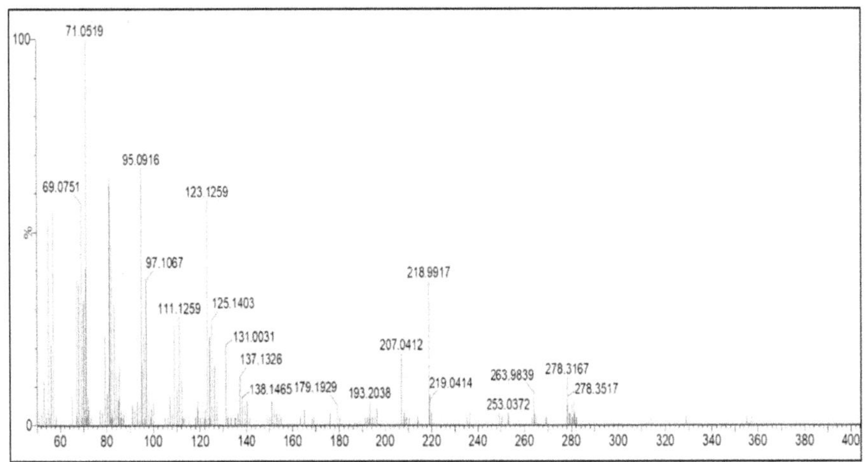

Figure 4.4: Mass spectrum of Isophytol

Quantification of aphrodisiac pheromones by GC-MS - The concentration of compounds A, B, C, and D (figure 4.2) in the extracts was calculated using integrated GC-MS peaks. They were calculated by comparing n-pentadecane integration as an internal standard in table 4.1. In males, the amounts of A, B, and C were 5922.48 ng/µl, 58.04 ng/µl, and 45.12 ng/µl respectively. In females, the amounts of A, B, and C were 777.38 ng/µl, 29.46 ng/µl, and 8.36ng/µl respectively in one butterfly. Hexahydrofarnesylacetone (B) and isophytol (C) were found in larger amounts in males than females, but this was not significant. The amount of E-phytol (A) in males was ten-fold higher than in females. The amount of brassicalactone (D) in males was 25.45 ng/µl, while in females it was not found. The graph (figure 4.4) showed the increased amounts of compounds A, B, and C produced by one butterfly, in logarithmic values.

Table 4.1: Amounts of compounds in the wings of one butterfly.

No.	Compounds	Retention Time (min)	Retention Index	Amounts in male wings per butterfly (ng/µl)	Amounts in female wings per butterfly (ng/µl)
1.	n-pentadecane	8.742	1501	i.s.*	i.s.*
2.	Hexahydrofarnesylacetone	11.392	1847	58.04±5.01	29.46±3.10
3.	Isophytol	12.242	1948	45.12±10.47	8.36±1.92
4.	Brassicalactone	12.676	1998	25.45±25.90	0.00
5.	E-phytol	14.059	2117	5922.48±714.34	777.38±207.73

*i.s.=used as internal standard.

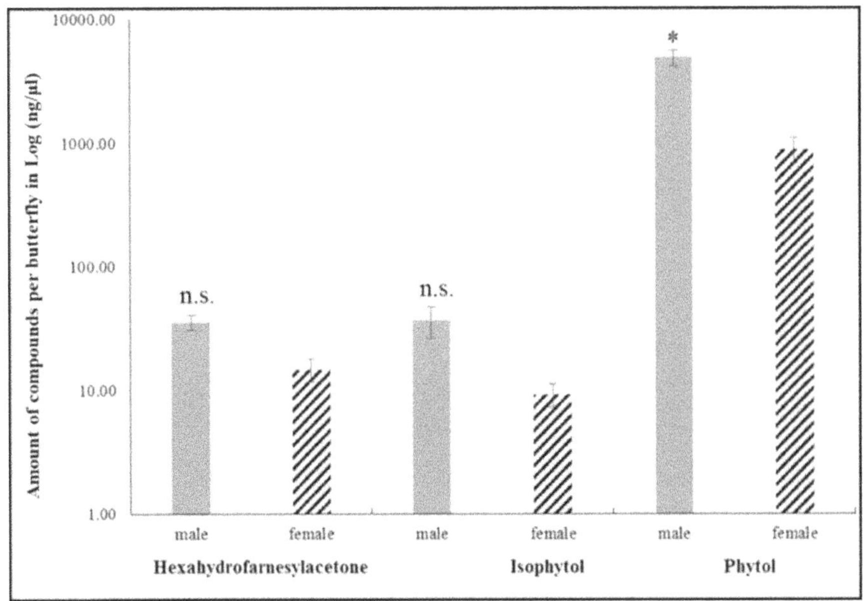

Figure 4.4: Amounts of pheromone compounds in wing extracts of one *P. brassicae* butterfly male or female. (* shows significance and n.s. shows non-significance between males and females, Fisher's LSD $P < 0.05$).

Electroantennographic response via puffing - Puffing was done using glass syringes. Hexane on the filter paper in the syringe was used as the negative control. Allyl isothiocyanate was used as a positive control as it was used as lures in trap experiments for various insects (Burgess and Wiens 1980; Pivnick et al. 1992). Male wings extract was also checked, and it showed a good signal (figure 4.5). We checked HHA, isophytol, and E-phytol separately. Capped female antenna were used to check the response. All samples were repeated ten to twelve times. Antennae of females responded to only hexahydrofarnesylacetone and E-phytol (figure 4.6). The average EAG response to these compounds separately was approximately half that of the male wings extract which was about 100mV. We did not get any EAG signal to isophytol. The different ratios of these three compounds were also puffed and checked, but on repetition, it did not show a satisfactory result.

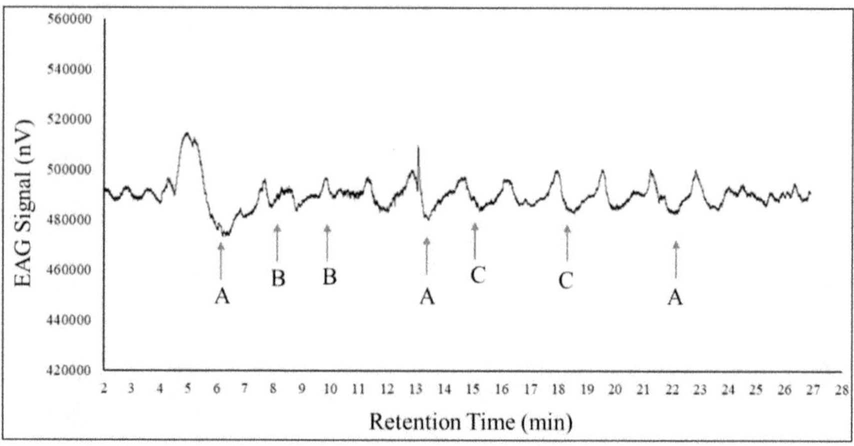

Figure 4.5: EAG response of *P. brassicae* female antennae (Germany, 6-8 days old) to the positive control allyl isothiocyanate (A), hexane as negative control (B) and male wings extract (C).

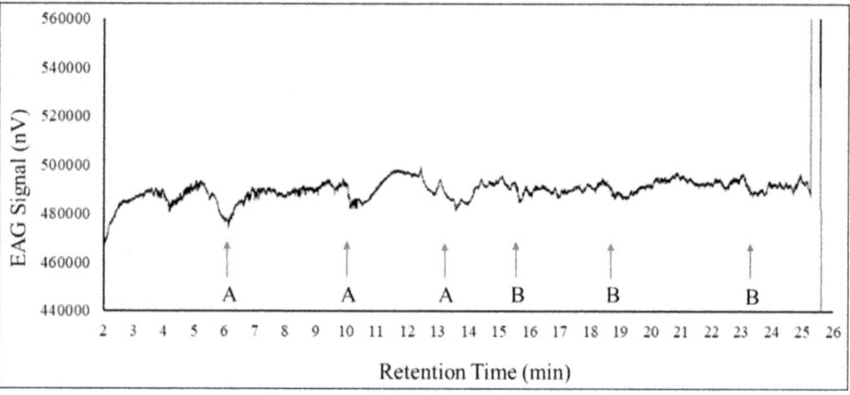

Figure 4.6: EAG response of *P. brassicae* female antennae (Germany, 6-8 days old) to the phytol (A) and HHA (B)

Behavioral bioassay - In wind tunnel experiment, male butterflies and the pheromones extracts of male wings were used. As the negative response was: 1) no flight of a butterfly, 2) the flight of butterfly away from the source of the pheromone, 3) fluttering with no direction throughout the wind tunnel, and 4) mate- refusal behavior (Ohguchi and Hidaka 1988; Obara 1964, Suzuki et al. 1977). A positive response was: 1) fluttering towards the pheromone source and settle down near the pheromone source (about 30cm), 2) pheromone source contact (when the female took a flight towards the pheromone source and its wings touch the source), and 3) females assuming a rigid posture and extends her abdomen out from the hindwings showing the male acceptance behavior (Rutowski 1976). We have found a negative response for all pure compounds at a dosage of 0.2µg. It means sometimes the butterflies were flying here and there and sometimes they sit at one position in the tunnel opposite to the pheromone source during the experiment. Two out of ten also showed mate-refusal behavior for pure HHA, which was also considered a negative response. The ratios 1:1:10, 1:1:20, 1:1:30, and 1:1:70 of HHA, Isophytol, and Phytol were used to check the response of females and when these mixtures of different compounds were applied, positive responses of females were observed. Most of the females have taken flight within 5 minutes after releasing them from their origin and settle them down in the vicinity of the pheromone source. Only a few females tried to come in contact with the pheromone source. With the mixture 1:1:70, females showed a 93.75% positive response which is similar to the response when the original pheromones extracts were used (Table 4.2). When the dummy was moved manually by waving the rod as though the male butterfly was flying, the patrolling virgin female usually approached. Within a 30 cm range of the dummy, individual females approaching randomly were scored for touching and landing on the specimen with extract during a test period of 20 min. The 1:1:70 solution induced a significant response as compared to the others (1:1:10 (HHA, I, P), 1:1:20 (HHA, I, P), 1:1:30 (HHA, I, P)). While the 1:1:30 (HHA, I, P) solution showed 40% of positivity as compared to the one with only hexane. It was better than the blank but statistically, it was not significant.

Table 4.2: Results of a behavioral assay showing the responses of female butterflies in response to male pheromone constituents

No.	Compounds to be tested	Total no. of female butterflies	Negative response (%)	Positive response (%)	Significance (Wilcoxon test, $p<0.05$)
1.	Only hexane	10	100	0	1.000
2.	100% phytol (P)	10	80	20	0.157
3.	100% hexahydrofarnesylacetone (HHA)	10	100	0	1.000
4.	100% isophytol (I)	10	100	0	1.000
5.	1:1:10 (HHA, I, P)	10	80	20	0.157
6.	1:1:20 (HHA, I, P)	10	90	10	0.317
7.	1:1:30 (HHA, I, P)	10	60	40	0.830
8.	1:1:70 (HHA, I, P)	14	21.43	78.57	0.001
9.	male pheromones extract	16	6.25	93.75	0.000
10.	live male butterfly	10	10	90.00	0.000

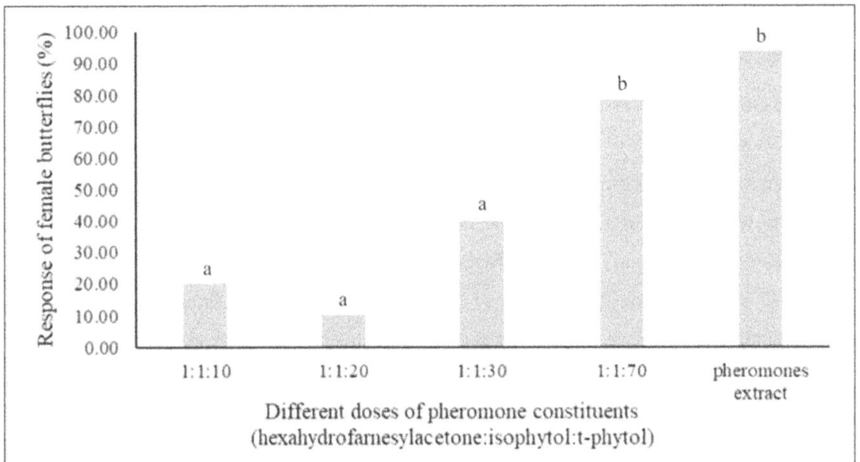

Figure 4.4: Response (%) of female butterflies to different doses of pheromone components. Different letters indicate significant differences between different doses of the same mixture of pheromone components (Wilcoxon test, $p<0.05$).

4.5 Discussion

Aphrodisiac pheromones are the chemical signals that are produced in the wings of male butterflies. They use these signals to attract females of the same species. We have found four compounds in the wings of male *P. brassicae*: brassicalactone, hexahydrofarnesylacetone (HHA), isophytol, and E-phytol in the German population, as they were documented by Yildizhan in the Netherlands population (2009).

According to our results, brassicalactone is only produced in male wings. While both male and female butterflies produce HHA, isophytol, and E-phytol but in different concentrations. Males produce these compounds in larger quantities than females. Our findings are in accordance with the research work of Yildizhan (2009). The quantitative difference between the male and female wings extracts helps in the process of identifying a suitable partner for courtship. HHA was also present in the hair pencil extracts of male *Tirathaba mundella* Walker, palm oil bunch moth) and the females use the presence of HHA to find a perfect partner for courtship (Sasaerila et al. 2003).

Another compound, E-phytol was produced in male *P. brassicae* with a quite big difference than in females, which is about ten-fold higher. E-phytol was also found to be a part of other insect wing compounds such as *Ephestia elutella*. A specific ratio of phytol with a γ-lactone, elicited a female response equivalent to that produced by the unfractionated wings extract (Phelan et al. 1986). One compound that is isophytol, an isomer of E-phytol was also found to be a part of both males and females wings pheromones of *P. brassicae*. In this study, isophytol was also considered to influence courtship. Therefore, it was a part of the bioassay as well as EAG studies.

Quantification of insect pheromones was carried out for decades. The main compounds of volatiles collected from calling males of great wax moth, *Galleria mellonella* were nonanal and undecanal and their ratio found in the Canadian population were 1: 3 (Romel et al. 1992) while in the USA population was 7: 3 (Leyrer and Monroe 1973). In this study, we have found 1:1.8:2.3:232.7; brassicalactone, isophytol, HHA, and E-phytol. Hence, we used different ratios of synthetic compounds in EAG studies. As the study of *G. mellonella* indicates that females' responses to the synthetic bait in laboratory tests were not as high as their response to alive males (Finn and Payne 1977) and in the field tests, the males were significantly more attractive than the pheromone baits tested. (Flint and Merkle 1983). Therefore, it is interesting to speculate the use of signaling substances and the information they convey.

In this study, we checked the activity of single compounds and their ratios that have been found in the wings of male *P. brassicae*. Therefore, we have carried out the electroantennographic studies, in which the response of female antenna to the single synthetic compounds was checked. The female antenna recognized the pure synthetic HHA, and E-phytol and gave a clear response as an EAG signal. Both compounds elicited the EAG response of approximately half amplitude as with the male wings extract. The females seem to reject the synthetic compound isophytol. In the study of aphrodisiac compounds of *P. brassicae*, the same EAG result was observed (Yildizhan et al. 2009).

To approve the activity of wings compounds singly as well as in specific mixtures, we carried out a wind-tunnel bioassay. In this bioassay, the response of female butterflies against pure synthetic HHA, isophytol, and E-phytol was observed. Some specific ratios of these three compounds were also used to get a specific combination of synthetic compounds, which could be later used as pheromone lures in field trials. The bioassay results reveal that the females are not attracted to the single pure compounds, but a mixture of three substances has been found to be active. For the single compounds, sometimes they did not even take a flight and sometimes they took a unidirectional flight overall in the wind tunnel, which was not considered as a positive response (Rutowski 1977a; 1977b). In the male wings extract, we have found E-phytol in larger quantities as HHA, and isophytol. Therefore, the ratios of compounds containing major proportion of E-phytol, were selected with HHA and isophytol as minor proportions. For the mixtures, the females were attracted to the ratios with 1:1:30 (HHA, isophytol, and phytol) and 1:1:70 only. A few of them were flattering around and resting very close to the dummy male with a 1:1:70 mixture. With the male wings extract, the females were behaving in the same way as with the ration mixture 1:1:70 (HHA, isophytol, and phytol). A few females were even observed touching the pheromone source having wings extract while flying from their origin. The females seem to reject the single wings compounds as they were behaving as if the wind tunnel did not have any pheromone source. Only two females showed mate-refusal posture (Suzuki et al. 1977) with pure HHA. It could be due to the unsuitable concentration of HHA, as the same behavior was observed by Ohguchi with the *Pieris melete* and *Pieris rapae* in a male recognition experiment. Both females took the mate-refusal posture and discontinued the posture only when they found a conspecific male (Ohguchi and Hidaka 1988). However, it can be concluded that the females accept the males during courtship based on the whole bouquet in a specific ratio that they release.

Female receptivity is usually signaled by a combination of visual and olfactory signals, visual stimuli being more important at a distance and olfactory stimuli being more important in close vicinity (Silberglied 1984). Aphrodisiac pheromones produced in the wings are used by the male butterflies to attract the females to bring them finally towards courtship. In the present study, we quantified these wings compounds of male *P. brassicae*, and the female response towards these male compounds was checked. As a result, we got a specific ratio of synthetic compounds. This ratio could be used as lures to distract the females in future field trials as well as in the agricultural fields of *Brassica* crops.

5. Host plant suitability of various *Brassica* cultivars for Pieris species and their response to herbivory

5.1 Abstract

A host plant screening was performed using two specialist herbivores, the European *Pieris rapae* and *Pieris brassicae*, and the induction of glucosinolates in *Brassica* varieties upon specialist herbivory was also investigated. For this experiment, Brussels sprouts (*Brassica oleracea* var. *gemmifera* DC. 'Igor'), Kale (*Brassica oleracea* var. *sabellica* L. 'Halbhoher Grüner Krauser'), Marrow-stem Kale (*Brassica oleracea* var. *medullosa* Thell. 'Grüner Ring'), Chinese cabbage (*Brassica rapa* ssp. *pekinensis* 'Kasumi F1'), and Savoy cabbage (*Brassica oleracea* var. *sabauda* L. "Vorbote/Hilmar) were selected. The second instar larvae of each species were allowed to feed on 4 weeks old plants (vegetative stage) for 7 days till they cross their fourth instar. Insects' performance of both specialists, measured as weight increase of larvae, was positively related to total GS levels of all varieties. The performance of *P. rapae* larvae did not show any significant difference in the host varieties, whereas *P. brassicae* larvae showed the best growth only on Marrow-stem Kale. Our GS analysis indicates that *P. rapae* feeding elicited the increase in aliphatic GS content only in Marrow-stem Kale, whereas *P. brassicae* herbivory showed an increase in Marrow-stem as well as in Savoy cabbage. The aliphatic GS levels did not change in all other varieties. The indolyl GS levels in all *Brassica* varieties were increased to several folds upon feeding of both specialist pests. All the *Brassica* varieties selected for our study had mainly 2-hydroxy-3-butenyl GS as a major component of aliphatic GS and the dominant compound of indolyl GS in all *Brassica* varieties was indol-3-ylmethyl GS. Both compounds showed major variations upon feeding damages.

Between the two indolyl GS, 4-methoxy-indolyl-3-methyl GS, and 1-methoxy-indolyl-3-methyl GS, the later one showed more elevation in concentration upon herbivory. With *P. rapae*, 1-methoxy-indolyl-3-methyl GS was increased about 30-fold in Savoy cabbage and Brussels sprouts, and about 20-fold in Marrow-stem Kale and Kale. With *P. brassicae*, 1-methoxy-indolyl-3-methyl GS was increased about 30-fold in Brussels sprouts while in Savoy cabbage, Kale, and Marrow-stem Kale about ten-fold increase was observed. In Chinese cabbage, both specialists induced more the induction of 4-methoxy-indolyl-3-methyl GS than 1-methoxy-indolyl-3-methyl GS as compared to the controls. These results were thought to use to find a correlation between specific GS and larval performance of the two specialist species, but we did not find any significant correlation between them.

5.2 Introduction

Brassicaceae is comprised of plants of wide economic importance that are used, for example as spices, oils, vegetables, and pharmaceuticals. The consumption of Brassicaceae vegetables containing Glucosinolates (GS), plant secondary metabolites might reduce the risk of carcinogenesis and in particular, disease in humans (Traka and Mithen 2009; Prakash and Gupta 2012; Melchini et al. 2013). GS has a characteristic chemical structure with a sulfur-linked ß-D-glucopyranose moiety and an amino acid-derived side chain. They are further divided based on this side chain into three groups, aliphatic, aromatic, and indole GS. The GS of Brassica plants has been studied for decades (Kjaer 1976; Feeney et al. 1970). More than 140 GS have been identified, 30 of which are present in Brassica species (Fahey et al. 2001; EFSA 2008; Bellostas et al. 2007).

The cabbage white butterflies, *P. rapae*, and *P. brassicae* are specialized in the plants of the family Brassicaceae. The *Pieris* adults exploiting the presence of GS in *Brassica* plants during the selection of host plants for oviposition (Renwick et al. 1992; van Loon et al. 1992). Therefore, *P. brassicae* females laid eggs on nonhost substrates dipped in GS solutions of host plants (David and Gardiner 1962). While their caterpillars use the presence of GS during the selection of host plants for feeding (Moyes et al. 2000). The *P. brassicae* caterpillars of the late-second instar until the end of the fifth instar consume leaves, buds, and flowers of *Brassica* plants (Smallegange et al. 2007).

The role of GS in host plants is known to mediate the herbivore-plant association. GS does not seem to be toxic themselves but when they are brought together with myrosinase (ß-thioglucosidases; E.C. 3.2.1.147), they are rapidly hydrolyzed to toxic isothiocyanates and other biologically active products and some other compounds such as nitriles, thiocyanates, epithionitriles, and oxazolidine thiones (Fenwick et al. 1983; Halkier and Gershenzon 2006; Kim and Jander 2007; Ahuja et al. 2010). GS levels have been demonstrated to be alerted due to the damage caused by several insect pests (Koritsas et al. 1991; Bodnaryk 1992; Hopkins et al. 1998; 2009). The GS concentration can be increased in response to herbivore feeding and this increase can affect the performance of specialist and generalist herbivores (Rask et al. 2000; Agarwal and Kurashige 2003; Gols et al. 2008; Gols and Harvey 2009). Likewise, the feeding of *Mamestra brassicae* L. on *B. napus* increased the levels of indol-3-yl-methyl, 1-methoxy-indol-3-yl-methyl, and total GS content in the plants (Ahuja et al. 2015). *Phyllotreta striolata* F. flea beetles use host plants' GS differently. They utilize GS to create their glucosinolates-myrosinase system. They accumulate up to 1.75% of their body weight in GS from their food plants and express their myrosinases to release alkenyl isothiocyanates (Beran et al. 2014). Phloem-feeding aphids (generalist *Myzus persicae* and specialist *Brevicoryne brassicae*) induced GS accumulation, particularly in short-chain aliphatic methylsulfinyl GS in *Arabidopsis thaliana* (Mewis et al. 2005). However, the caterpillar of *P. rapae* did not cause a significant induction of aliphatic GS, but indolyl GS contents, mainly 1-methoxyindol-3-ylmethyl GS increased significantly in *A. thaliana* (Columbia wild-type) (Mewis et al. 2006).

With this background, this study aimed to investigate the insect performance of two specialist herbivores, *P. rapae*, and *P. brassicae*, when the larvae were forced to feed on one host plant. Furthermore, we investigated the GS accumulation in response to herbivory in five cultivars of *Brassica* plants. We hypothesized that the herbivory of a specific specialist has a significant impact on the concentration of some specific GS component.

5.3 Materials and Methods

Insects - Larvae from *P. rapae* and *P. brassicae* were collected from the open trap fields at the Julius Kühn-Institut, Berlin and then reared as described in chapter 2.

Plant material - For this experiment Brussels sprouts (*Brassica oleracea* var. *gemmifera* DC. 'Igor'), Kale (*Brassica oleracea* var. *sabellica* L. 'Halbhoher Grüner Krauser'), Marrow-stem Kale (*Brassica oleracea* var. *medullosa* Thell. 'Grüner Ring'), Chinese cabbage (*Brassica rapa* ssp. *pekinensis* 'Kasumi F1'), and Savoy cabbage (*Brassica oleracea* var. *sabauda* L. 'Vorbote/Hilmar') were selected as host plants. First, seeds were sown in a tray with Jungpflanzensubstrat S1 from Klasmann-Deilmann, Germany. They were placed in a Green house chamber at 24°C. Then small plants were picked and transferred from tray into small pots after one week. All plants used for the growth bioassay were about 4 weeks old. For the whole experiment, we need 30 plants per variety (10 plants per harvest).

Forced-feeding experiment - The experiment was carried out with *P. rapae* and *P. brassicae* larvae, with one individual per plant. The whole set-up was in a climate chamber with 16 h photoperiod at 22°C to 24°C. The host plant suitability tests were done with seven replications for each plant variety. Three plants were not treated with caterpillars and served as control. Before starting the experiment, larvae were weighted. The old plants were replaced twice with the new ones to provide enough food harvested, first after three days and secondly, after two days. The plants were harvested two times in liquid Nitrogen for the GS extraction later, first after two days (on day 5, H1), and again at the end of the experiment (on day 7H2). After seven days, the experiment was completed, and the final larvae weight was determined to calculate the weight gain throughout the experiment. The harvested plants were used for GS extractions.

Extraction of GS from plant leaves - All samples from harvested plants were first freeze-dried for 4 days at 0.03 mbar and 20°C. The samples were then pulverized in a grinding mill with the help of stainless-steel balls. An amount of 40 mg of the powdered leafy material was extracted using (750µl) 70% HPLC grade methanol. 1mM Sinalbin was used as internal standard (60µl). Extracts were centrifuged at 10,000 rpm for 5 min. The supernatants were separated. Extraction was repeated twice with (500µl) 70% HPLC grade methanol of and supernatants were combined. Extracts were desulphated on DEAE Sefadex mini column in 2M acetic acid. DEAE Sefadex mini columns were prepared by adding 500µl (1:3) Sefadex in acetic acid suspension. Columns were then rinsed twice with 1ml of 6M Imidazole-formate solution (in 30% Formic acid, Merck), 1ml MilliQ water. Extracts were then poured on the columns. Afterwards, the columns were washed twice with 1ml Sodium acetate buffer, pH 4.0. After drying the columns on tissue paper, 75µl Sulphatase from Helix pomatia (Sigma-Aldrich) was added to desulfate the extracts. Columns were capped and left overnight. Desulfated extracts were eluted after 24 h with 1ml of MilliQ water in HPLC sample vails.

HPLC analysis - The extracts after desulfation were separated and analysed by HPLC fitted with a 2.1*100 mm Poroshell 120EC-C18 reverse-phase column (2.7µm, Agilent USA). A gradient system of deionized Milli-Q water (solvent A) and HPLC grade acetonitrile (Solvent B) was used. The eluent system was as follows: 0-14 min: 0.5% B, 15-16 min: 100% B, 17-21 min: 0.5% B at a flow rate of 0.4 ml/min. The eluent was monitored by photodiode array detection between 190 nm and 400 nm with RS DAD (DIONEX Ultimate 3000, Thermo Scientific).

Statistical analysis of data - The statistical significance of variation in the amounts of aphrodisiac compounds among different populations was determined by using univariate generalized linear model (GLM), followed by posthoc test and Fischer`s least significant difference (LSD). All analysis was performed in SPSS (IBM SPSS Statistics 23).

5.4 Results

Forced feeding Bioassay

Insect performance of P. rapae and P. brassicae larvae on host plants - In this forced feeding bioassay, larvae were forced to feed on one variety of *Brassica* plants for 7 days. At the end of experiment, the larval weight was measured. All *P. rapae* larvae showed an increase in weight gain of about 3000% to 4000%. Feeding on Marrow-stem Kale and Savoy cabbage showed a slight better growth of *P. rapae* larvae. The statistical analysis by Tukey's test showed non-significant difference among all *Brassica* varieties. (figure 5.1). The same bioassay was carried out with *P. brassicae* larvae too. After the seventh day, the larval weight showed an increase between the range of about 4000% to 7000% as compared to the larval weight measured before starting the bioassay. Feeding on Marrow-stem Kale resulted in the highest weight gain of larvae. Secondly, feeding on Kale was better than the other varieties (figure 5.2).

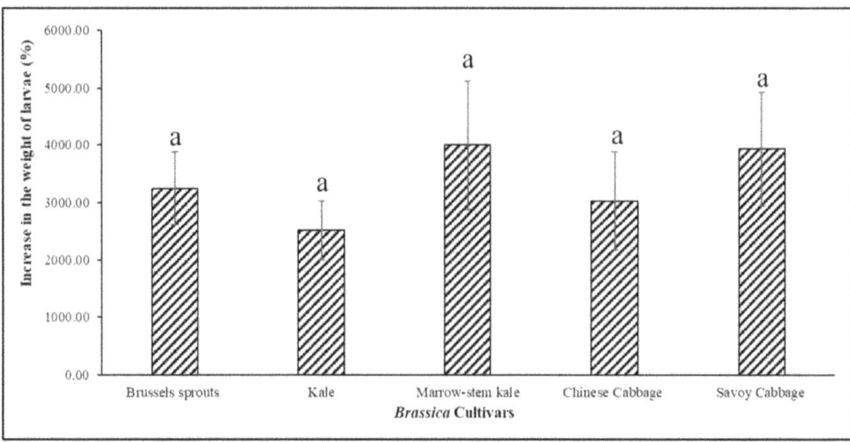

Figure 5.1: Increase in larval weight of *P. rapae* after 7 days of forced feeding bioassay on different crop varieties (Significant differences among cultivars calculated by posthoc and Tukey's test; $p \leq 0.05$, represented as letter a).

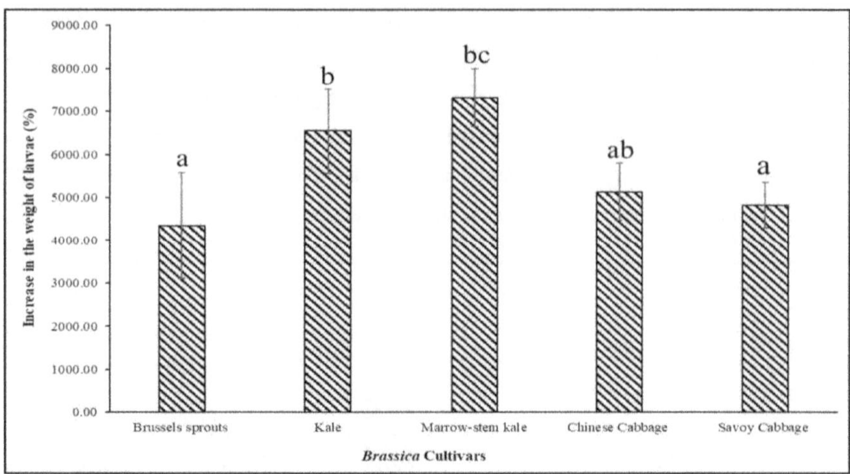

Figure 5.2: Increase in larval weight of *P. brassicae* after 7 days of forced feeding bioassay on different crop varieties (Significant differences among cultivars calculated by posthoc and Tukey's test; $p \leq 0.05$).

GS profiles in Host plants – Five *Brassica* varieties were used in the forced feeding experiment. HPLC analysis of GS contents in the leaves of harvested plants showed four aliphatic GS, 2-hydroxy-3-butenyl- (progoitrin), 2-propenyl- (Sinigrin), 4-methylsulfinylbutyl (glucoraphanin) and 3-butenyl (gluconapin), while the indolic GS include 4-hydroxyindol-3-ylmethyl (4-hydroxyglucobrassicin), indol-3-ylmethyl (glucobrassicin), 4-methoxyindol-3-ylmethyl (4-methoxyglucobrassicin) and 1-methoxyindol-3-ylmethyl (neoglucobrassicin). Four *B. oleracea* varieties contained 2-hydroxy-3-butenyl-GS as a major aliphatic GS and as major indolic GS, indol-3-ylmethyl-GS was found. The variety of *B. rapa*, Chinese cabbage does not contain detectable aliphatic GS levels, only the indolic GS were found, and the major component was 1-methoxyindol-3-ylmethyl-GS. Quantitative GS profiles for each variety were analyzed to be different for each variety before and after feeding with insects. These results are calculated as shown in tables 5.1-5.4 and summarized in figures 5.3-5.13.

Changes in the GS profile of plants (H1) upon P. rapae and P. brassicae herbivory - HPLC analysis of GS content of different varieties of *Brassica* plants was done. It revealed some significant variations in the GS profiles after 2 days of larvae feeding. At this time of experiment, larvae were between L3-L4 instar. All GS concentrations were described after comparing them to the GS levels in control plants. As a result of *P. rapae* herbivory, the level of aliphatic GS found to be significantly increased only in Marrow-stem Kale when compared to the plants without insect damage. The amount of aliphatic GS in Marrow-stem Kale was 9.77 µmol/g DW, which is1.78-fold increase than control plants. The increase in the amounts of aliphatic GS in all other *Brassica* varieties upon *P. rapae* herbivory was not significant and determined to be as follows; Brussels sprouts (5.54 µmol/g DW), Savoy cabbage (4.23 µmol/g DW) and Kale (2.55 µmol/g DW). Whereas upon *P. brassicae* herbivory, the level of aliphatic GS was found to be significantly increased only in Brussels sprouts (12.55 µmol/g DW) which was 14.21-fold increase, when compared to the plants without insect damage. The increase in the amounts of aliphatic GS in all other *Brassica* varieties upon *P. brassicae* herbivory was not significant and found to be as follows; Savoy cabbage (2.85 µmol/g DW), Marrow-stem Kale (2.34 µmol/g DW) and Kale showed no change in aliphatic GS.

The decrease in the amounts of indolic GS in Marrow-stem Kale, Chinese cabbage and Savoy cabbage upon *P. rapae* herbivory was significant. While the levels of indolic GS in Brussels sprouts and Kale showed a very slight decrease than the control plants. Upon *P. brassicae herbivory*, the level of indolic GS increased significantly in Chinese cabbage (4.16 µmol/g DW), Savoy cabbage (3.60 µmol/g DW) and Kale (2.56 µmol/g DW). The indolic GS content in Marrow-stem Kale did not show any change. While in Brussel sprouts, indolic GS decreased, as compared to the control plants.

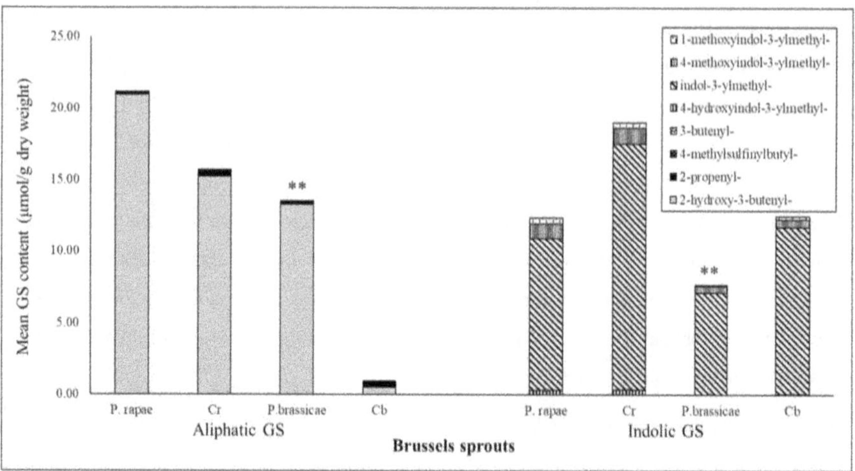

Figure 5.3: Aliphatic and indolic GS content of Brussels sprouts after two days feeding of L3-L4 instar *P. rapae* and *P. brassicae.* Cr represents control plants for *P. rapae* experiment and Cb represents control plants for *P. brassicae* experiment (* represents significant differences between plants affected upon herbivory and control plants different treatments within one variety, Fisher's LSD test; **$p \leq 0.05$; *$p < 0.1$).

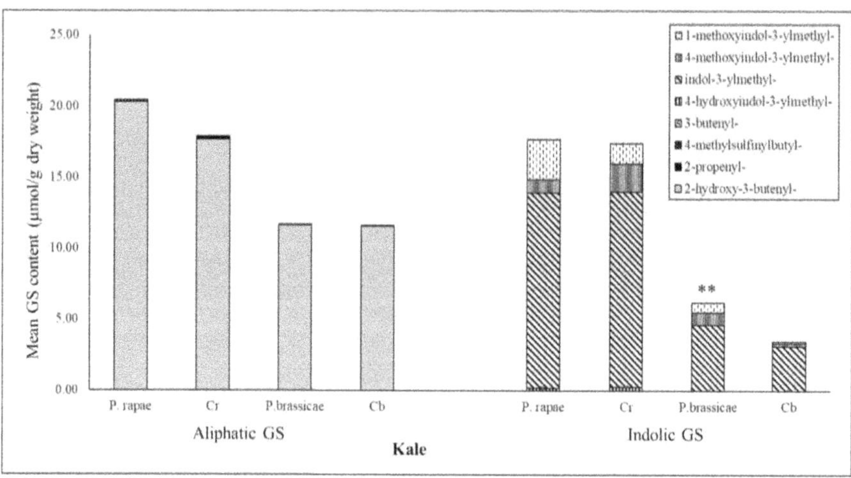

Figure 5.4: Aliphatic and indolic GS content of Kale after two days feeding of L3-L4 instar *P. rapae* and *P. brassicae*. Cr represents control plants for *P. rapae* experiment and Cb represents control plants for *P. brassicae* experiment (* represents significant differences between plants affected upon herbivory and control plants different treatments within one variety, Fisher's LSD test; **$p \leq 0.05$; *$p < 0.1$).

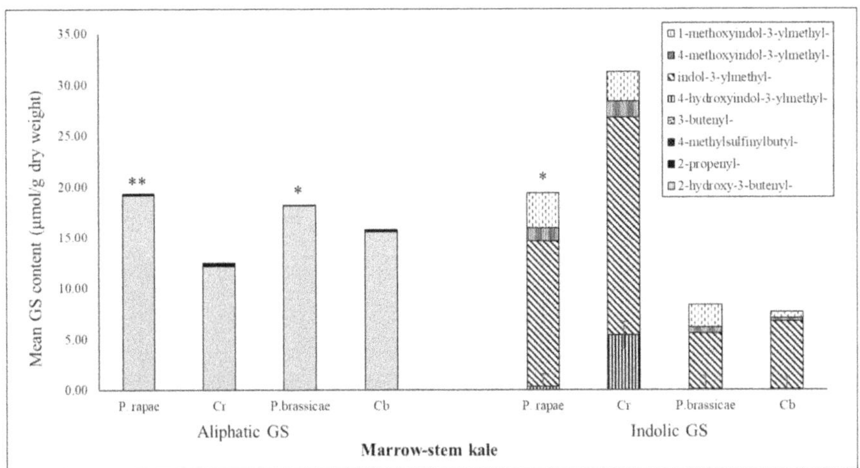

Figure 5.5: Aliphatic and indolic GS content of Marrow-stem Kale after two days feeding of L3-L4 instar *P. rapae* and *P. brassicae*. Cr represents control plants for *P. rapae* experiment and Cb represents control plants for *P. brassicae* experiment (* represents significant differences between plants affected upon herbivory and control plants different treatments within one variety, Fisher's LSD test; **$p \leq 0.05$; *$p < 0.1$).

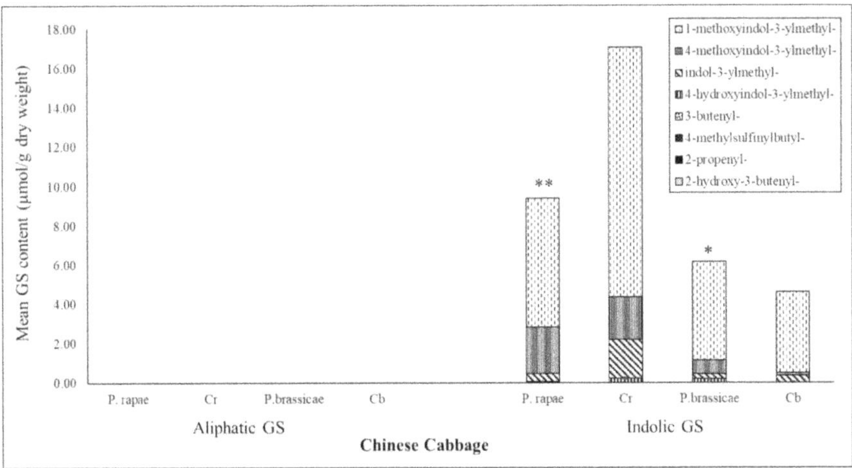

Figure 5.6: Aliphatic and indolic GS content of Chinese cabbage after two days feeding of L3-L4 instar *P. rapae* and *P. brassicae*. Cr represents control plants for *P. rapae* experiment and Cb represents control plants for *P. brassicae* experiment (* represents significant differences between plants affected upon herbivory and control plants different treatments within one variety, Fisher's LSD test; **$p \leq 0.05$; *$p < 0.1$).

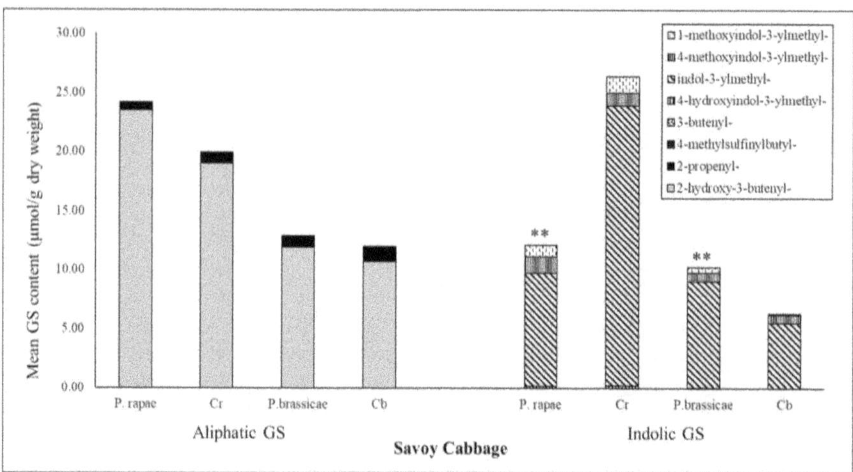

Figure 5.7: Aliphatic and indolic GS content of Savoy cabbage after two days feeding of L3-L4 instar *P. rapae* and *P. brassicae*. Cr represents control plants for *P. rapae* experiment and Cb represents control plants for *P. brassicae* experiment (* represents significant differences between plants affected upon herbivory and control plants different treatments within one variety, Fisher's LSD test; $**p \leq 0.05$; $*p < 0.1$).

Changes in the GS profile of plants (H2) upon P. rapae and P. brassicae herbivory - After 2 days, the GS contents (H2) were checked again when the larvae were between L4-L5 instar at the end of the experiment. As a result of *P. rapae* herbivory, the level of aliphatic GS found to be significantly increased only in Marrow-stem Kale (15.76 µmol/g DW) when compared to the plants without insect damage. The aliphatic GS increased very slightly in Brussels sprouts, and Savoy cabbage, while in Kale, *P. rapae* herbivory showed no effect. Upon *P. brassicae* herbivory, the highest significant increase in aliphatic GS content was determined to be in Savoy cabbage (12.78 µmol/g DW) which is 2.23-fold higher than the control plants. The other varieties showed either decreased or no effect in aliphatic GS levels.

The amount of indolic GS was significantly increased from 2.47 to 7.69 folds in varieties fed by the *P. rapae* larvae. The highest increase in indolic GS content in plants after larval feeding was determined to be in Kale (19.28 µmol/g DW) followed by Marrow-stem Kale (17.70 µmol/g DW), then in Savoy cabbage (11.82 µmol/g DW) followed by Brussels sprouts (9.22 µmol/g DW) (Table 5.2). Whereas the significant increase in indolic GS content was found in all four varieties of *Brassica* except Chinese cabbage. The highest increase was found to 3.76-fold higher in Kale (7.96 µmol/g DW), followed by Savoy cabbage (6.94 µmol/g DW), 2.36-fold increase in Marrow-stem Kale (8.31 µmol/g DW) and 1.80-fold higher in Brussels sprouts (4.98 µmol/g DW). When the total GS amounts were considered, the herbivory of both specialists elicited significant increase in all *Brassica* varieties. Upon damage with *P. rapae* larvae, the increase in total GS levels was from 1.55 to 3.12-folds higher and in *P. brassicae* damaged plants, the total GS levels were from 1.19 to 2.23-folds higher as compared to the controls.

Induction of specific GS due to larval feeding – Savoy cabbage contains all four aliphatic and four indolic GS in suitable amounts, while Chinese cabbage contains only four indolic GS. the herbivory of both specialists elicited the aliphatic GS levels in some *Brassica* varieties, which was mainly due to an increase of 2-hydroxy-3-butenyl GS and the increase in indolic GS was due to Indol-3-ylmethyl-GS. Induction of 4-Methoxyindol-3-yl methyl-GS and 1-Methoxyindol-3-yl methyl-GS were also found in significant higher amounts in the herbivory affected plants. (figure 5.7 and 5.8).

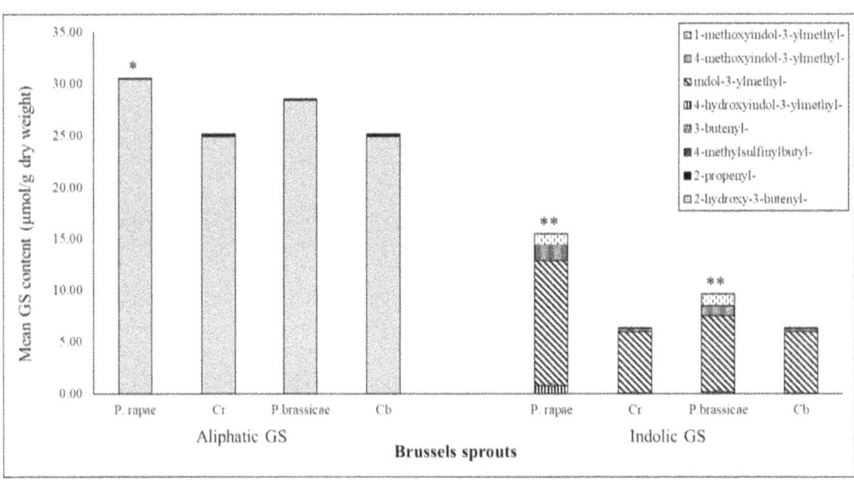

Figure 5.8: Aliphatic and indolic GS content of Brussels sprouts after two days feeding of L4-L5 instar *P. rapae* and *P. brassicae*. Cr represents control plants for *P. rapae* experiment and Cb represents control plants for *P. brassicae* experiment (* represents significant differences between plants affected upon herbivory and control plants different treatments within one variety, Fisher's LSD test; $^{**}p \leq 0.05$; $^{*}p < 0.1$).

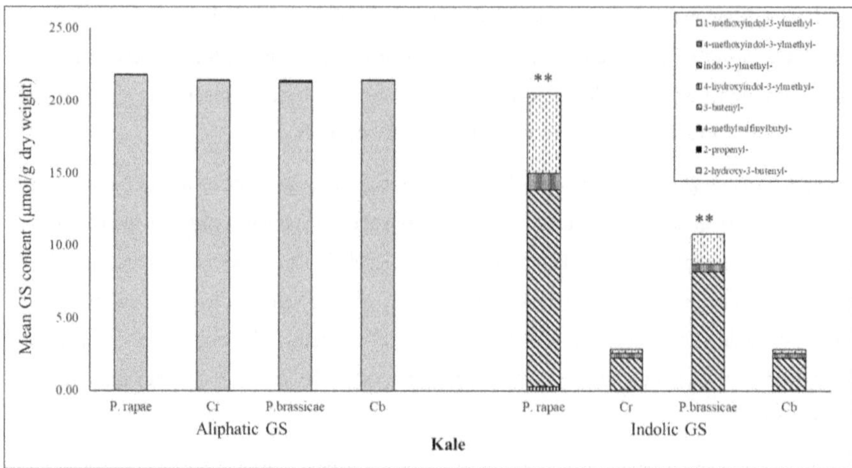

Figure 5.9: Aliphatic and indolic GS content of Kale after two days feeding of L4-L5 instar *P. rapae* and *P. brassicae*. Cr represents control plants for *P. rapae* experiment and Cb represents control plants for *P. brassicae* experiment (* represents significant differences between plants affected upon herbivory and control plants different treatments within one variety, Fisher's LSD test; $^{**}p \leq 0.05$; $^{*}p < 0.1$).

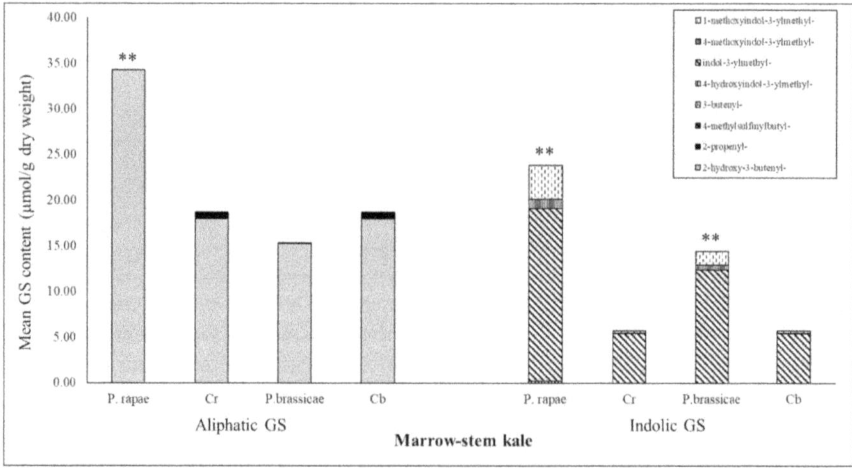

Figure 5.10: Aliphatic and indolic GS content of Marrow-stem Kale after two days feeding of L4-L5 instar *P. rapae* and *P. brassicae*. Cr represents control plants for *P. rapae* experiment and Cb represents control plants for *P. brassicae* experiment (* represents significant differences between plants affected upon herbivory and control plants different treatments within one variety, Fisher's LSD test; $^{**}p \leq 0.05$; $^{*}p < 0.1$).

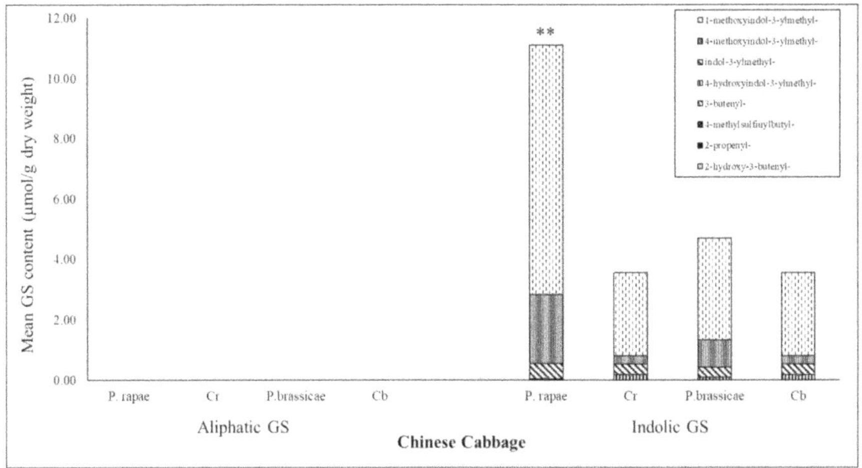

Figure 5.11: Aliphatic and indolic GS content of Chinese cabbage after two days feeding of L4-L5 instar *P. rapae* and *P. brassicae.* Cr represents control plants for *P. rapae* experiment and Cb represents control plants for *P. brassicae* experiment (* represents significant differences between plants affected upon herbivory and control plants different treatments within one variety, Fisher's LSD test; $**p \leq 0.05$; $*p < 0.1$).

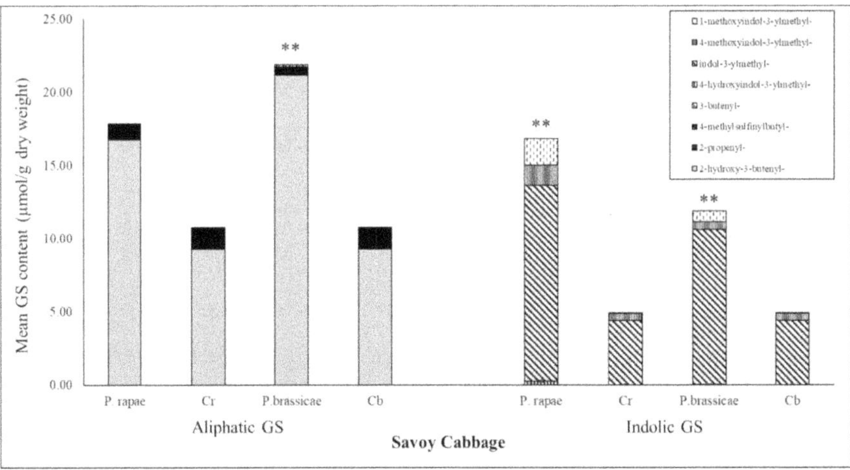

Figure 5.12: Aliphatic and indolic GS content of Savoy cabbage after two days feeding of L4-L5 instar *P. rapae* and *P. brassicae.* Cr represents control plants for *P. rapae* experiment and Cb represents control plants for *P. brassicae* experiment (* represents significant differences between plants affected upon herbivory and control plants different treatments within one variety, Fisher's LSD test; $**p \leq 0.05$; $*p < 0.1$).

Relationship between Insect Performance and GS contents - We calculated the Pearson product-moment correlations to check the relationship between plant GS levels and weight gain of larvae. The Pearson coefficient for *P. rapae* was - 0.148 and for *P. brassicae* was - 0.229. Both species' larval weight gain was negative in response to GS contents. Therefore, it was concluded that the insects weight gain was not related to GS levels found in different varieties of plants. Correlation between each GS type and larval weight gain was calculated, and we found all the correlations to be non-significant (table 5.6 and 5.7).

Table 5.1: GS contents found in different *Brassica* varieties upon L3-L4 instar *P. rapae* herbivory. Treated showed the GS amounts in the plants after feeding and Control showed the GS amounts in the plants without feeding.

No.	Chemical names of GS	Rt (min)	Occurrence of GS in various cultivars of cabbage									
			Brussel sprout		Kale		Marrow-stem Kale		Chinese cabbage		Savoy cabbage	
			Treated	Control	Treated	Control	Treated	Control	Treated	Control	Treated	Control
1.	2-hydroxy-3-butenyl GS	1.85	20.97 ±4.49	15.26 ±5.78	20.33 ±3.44	17.72 ±7.07	19.12 ±7.91	12.13 ±1.67	0.00	0.00	23.51 ±2.88	19.04 ±8,51
2.	2-propenyl GS	3.02	0.15 ±0.06	0.40 ±0.05	0.02 ±0.03	0.11 ±0.06	0.18 ±0.12	0.25 ±0.02	0.00	0.00	0.59 ±0.09	0.82 ±0.10
3.	4-Methyl sulfinylbutyl GS	3.64	0.02 ±0.02	0.07 ±0.02	0.10 ±0.03	0.08 ±0.01	0.00	0.06 ±0.01	0.00	0.00	0.00	0.02 ±0.03
4.	3-Butenyl GS	6.73	0.00	0.00	0.00	0.00	0.00	0.00	0.00	0.00	0.07 ±0.03	0.06 ±0.00
5.	4-Hydroxy indol-3-ylmethyl GS	7.37	0.32 ±0.16	0.34 ±0.07	0.21 ±0.07	0.30 ±0.02	0.29 ±0.05	5.37 ±0.03	0.40 ±0.01	0.22 ±0.04	0.19 ±0.06	0.24 ±0.12
6.	indol-3-ylmethyl GS	9.40	10.65 ±5.22	17.22 ±0.26	13.73 ±2.79	13.74 ±1.05	14.30 ±7.37	21.38 ±0.42	0.42 ±0.02	1.96 ±0.18	9.53 ±2.43	23.69 ±2.65
7.	4-methoxyindol-3-ylmethyl GS	10.30	1.02 ±0.03	1.08 ±0.02	0.92 ±0.12	1.94 ±0.11	1.32 ±0.04	1.59 ±0.01	2.36 ±0.19	2.16 ±0.14	1.38 ±0.26	1.11 ±0.21
8.	1-methoxyindol-3-ylmethyl GS	11.76	0.40 ±0.22	0.40 ±0.01	2.84 ±0.22	1.46 ±0.05	3.47 ±0.80	2.88 ±0.20	6.56 ±1.99	12.72 ±0.54	0.98 ±0.14	1.39 ±0.39
	Aliphatic GS		21.27 ±4.99	15.73 ±5.84	20.46 ±3.47	17.91 ±7.04	22.22 ±7.88	12.45 ±1.68	0.00	0.00	24.17 ±2.91	19.94 ±8.63
	Indolic GS		14.60 ±0.47	19.04 ±0.20	17.70 ±2.95	17.43 ±1.18	21.47 ±7.74	26.26 ±0.53	9.37 ±2.16	16.66 ±0.84	13.34 ±0.23	26.74 ±2.44
	Total GS		35.87 ±4.69	34.77 ±5.85	38.15 ±5.79	35.34 ±8.17	44.66 ±3.37	38.47 ±1.54	9.37 ±2.16	16.66 ±0.84	36.26 ±3.28	33.45 ±6.27

Table 5.2: GS contents found in different *Brassica* varieties upon L4-L5 instar *P. rapae* herbivory. Treated showed the GS amounts in the plants after feeding and Control showed the GS amounts in the plants without feeding

No.	Chemical names of GS	Rt (min)	Occurrence of GS in various cultivars of cabbage									
			Brussel sprout		Kale		Marrow-stem Kale		Chinese cabbage		Savoy cabbage	
			Treated	Control	Treated	Control	Treated	Control	Treated	Control	Treated	Control
1.	2-hydroxy-3-butenyl GS	1.85	30.43 ±7.12	24.90 ±2.10	21.79 ±7.10	21.39 ±0.71	34.29 ±3.42	17.98 ±2.92	0.00	0.00	16.74 ±2.46	9.28 ±2.23
2.	2-propenyl GS	3.02	0.03 ±0.03	0.15 ±0.09	0.03 ±0.02	0.00	0.00	0.65 ±0.29	0.00	0.00	1.04 ±0.03	1.43 ±0.22
3.	4-Methylsulfinyl butyl GS	3.64	0.00	0.01 ±0.01	0.00	0.06 ±0.03	0.00	0.06 ±0.04	0.00	0.00	0.00	0.03 ±0.20
4.	3-Butenyl GS	6.73	0.00	0.00	0.00	0.00	0.00	0.00	0.00	0.00	0.07 ±0.01	0.04 ±0.01
5.	4-Hydroxyindol-3-ylmethyl GS	7.37	0.78 ±0.07	0.02 ±0.01	0.29 ±0.09	0.02 ±0.02	0.26 ±0.03	0.02 ±0.07	0.05 ±0.04	0.18 ±0.09	0.26 ±0.05	0.00
6.	indol-3-ylmethyl GS	9.40	12.05 ±1.17	5.97 ±0.11	13.59 ±3.67	2.28 ±0.10	18.89 ±0.74	5.39 ±0.41	0.51 ±0.05	0.37 ±0.04	13.38 ±0.42	4.37 ±0.30
7.	4-methoxyindol-3-ylmethyl GS	10.30	1.55 ±0.16	0.34 ±0.01	1.16 ±0.30	0.33 ±0.02	1.00 ±0.02	0.10 ±0.01	2.28 ±0.17	0.27 ±0.08	1.36 ±0.08	0.50 ±0.03
8.	1-methoxyindol-3-ylmethyl GS	11.76	1.09 ±0.14	0.03 ±0.01	5.51 ±1.17	0.25 ±0.01	3.67 ±0.27	0.19 ±0.04	8.27 ±0.38	2.74 ±0.24	1.83 ±0.38	0.05 ±0.00
	Aliphatic GS		32.86 ±4.53	25.06 ±2.04	20.45 ±2.61	21.47 ±0.75	34.29 ±3.42	18.53 ±2.64	0.00	0.00	17.86 ±2.47	11.82 ±1.21
	Indolic GS		15.48 ±1.51	6.26 ±0.17	22.16 ±2.81	2.88 ±0.13	23.80 ±	6.10 ±0.57	11.11 ±0.49	3.56 ±0.31	16.78 ±0.68	4.96 ±0.30
	Total GS		48.59 ±5.54	31.42 ±1.93	43.69 ±0.30	24.35 ±0.71	58.11 ±3.10	24.50 ±3.03	11.11 ±0.49	3.56 ±0.31	34.69 ±1.79	16.88 ±1.03

Table 5.3: GS contents found in different *Brassica* varieties upon L3-L4 instar *P. brassicae* herbivory. Treated showed the GS amounts in the plants after feeding and Control showed the GS amounts in the plants without feeding

No.	Chemical names of GS	Rt (min)	Occurrence of GS in various cultivars of cabbage									
			Brussel sprout		Kale		Marrow-stem Kale		Chinese cabbage		Savoy cabbage	
			Treated	Control	Treated	Control	Treated	Control	Treated	Control	Treated	Control
1.	2-hydroxy-3-butenyl GS	1.85	13.30 ±1.34	0.51 ±0.12	11.68 ±2.27	11.59 ±1.60	18.13 ±0.87	15.57 ±0.87	0.00	0.00	11.90 ±2.30	10.73 ±3.32
2.	2-propenyl GS	3.02	0.19 ±0.05	0.38 ±0.00	0.01 ±0.02	0.00	0.02 ±0.03	0.23 ±0.01	0.00	0.00	0.92 ±0.10	1.10 ±0.03
3.	4-Methylsulfinylbutyl GS	3.64	0.02 ±0.00	0.07 ±0.01	0.00	0.06 ±0.00	0.00	0.00	0.00	0.00	0.02 ±0.01	0.04 ±0.00
4.	3-Butenyl GS	6.73	0.00	0.00	0.00	0.00	0.00	0.00	0.00	0.00	0.10 ±0.04	0.11 ±0.00
5.	4-Hydroxyindol-3-ylmethyl GS	7.37	0.01 ±0.00	0.05 ±0.05	0.02 ±0.01	0.02 ±0.00	0.04 ±0.03	0.03 ±0.02	0.18 ±0.18	0.01 ±0.00	0.05 ±0.03	0.00
6.	indol-3-ylmethyl GS	9.40	7.13 ±0.48	11.71 ±0.57	4.62 ±0.30	3.11 ±0.19	5.53 ±2.66	6.72 ±0.42	0.27 ±0.05	0.35 ±0.03	8.99 ±2.21	5.56 ±0.23
7.	4-methoxyindol-3-ylmethyl GS	10.30	0.45 ±0.03	0.52 ±0.01	0.88 ±0.48	0.23 ±0.01	0.57 ±0.08	0.31 ±0.02	0.68 ±0.35	0.14 ±0.03	0.77 ±0.06	0.66 ±0.04
8.	1-methoxyindol-3-ylmethyl GS	11.76	0.07 ±0.01	0.23 ±0.02	0.69 ±0.19	0.11 ±0.00	2.17 ±0.14	0.58 ±0.04	5.01 ±3.30	4.10 ±0.28	0.44 ±0.08	0.11 ±0.01
	Aliphatic GS		13.50 ±1.38	0.95 ±0.13	10.73 ±0.90	11.65 ±1.61	18.14 ±0.88	15.80 ±0.86	0.00	0.00	12.93 ±2.37	10.08 ±0.31
	Indolic GS		7.76 ±0.42	12.54 ±0.46	6.11 ±0.57	3.55 ±0.20	8,31 ±2.76	7.69 ±0.36	8.77 ±0.79	4.61 ±0.34	9.94 ±2.16	6.34 ±0.28
	Total GS		21.07 ±1.78	13.46 ±0.78	16.78 ±0.58	15.12 ±1.41	27.67 ±2.85	23.44 ±0.36	8.77 ±0.79	4.61 ±0.34	23.18 ±2.19	16.33 ±0.01

Table 5.4: GS contents found in different *Brassica* varieties upon L4-L5 instar *P. brassicae* herbivory. Treated showed the GS amounts in the plants after feeding and Control showed the GS amounts in the plants without feeding.

No.	Chemical names of GS	Rt (min)	Occurrence of GS in various cultivars of cabbage									
			Brussel sprout		Kale		Marrow-stem Kale		Chinese cabbage		Savoy cabbage	
			Treated	Control	Treated	Control	Treated	Control	Treated	Control	Treated	Control
1.	2-hydroxy-3-butenyl GS	1.85	28.38 ±4.95	24.90 ±2.10	21.29 ±2.87	21.39 ±0.71	15.24 ±5.98	17.98 ±2.92	0.00	0.00	21.18 ±4.83	9.28 ±2.23
2.	2-propenyl GS	3.02	0.11 ±0.01	0.15 ±0.09	0.02 ±0.04	0.00	0.08 ±0.05	0.65 ±0.29	0.00	0.00	0.56 ±0.18	1.43 ±0.22
3.	4-Methylsulfinyl butyl GS	3.64	0.00	0.01 ±0.01	0.09 ±0.06	0.06 ±0.03	0.00	0.06 ±0.04	0.00	0.00	0.00	0.03 ±0.20
4.	3-Butenyl GS	6.73	0.00	0.00	0.00	0.00	0.00	0.00	0.00	0.00	0.13 ±0.05	0.04 ±0.01
5.	4-Hydroxyindol-3-ylmethyl GS	7.37	0.15 ±0.08	0.02 ±0.01	0.01 ±0.01	0.02 ±0.02	0.01 ±0.01	0.02 ±0.07	0.09 ±0.12	0.18 ±0.09	0.02 ±0.02	0.00
6.	indol-3-ylmethyl GS	9.40	7.37 ±2.54	5.97 ±0.11	8.23 ±1.04	2.28 ±0.10	12.38 ±3.89	5.39 ±0.41	0.35 ±0.06	0.37 ±0.04	10.60 ±2.09	4.37 ±0.30
7.	4-methoxyindol-3-ylmethyl GS	10.30	0.97 ±0.26	0.34 ±0.01	0.51 ±0.05	0.33 ±0.02	0.55 ±0.04	0.10 ±0.01	0.91 ±0.11	0.27 ±0.08	0.51 ±0.09	0.50 ±0.03
8.	1-methoxyindol-3-ylmethyl GS	11.76	1.13 ±0.44	0.03 ±0.01	2.09 ±1.28	0.25 ±0.01	1.46 ±0.61	0.19 ±0.04	3.35 ±1.04	2.74 ±0.24	0.77 ±0.15	0.05 ±0.00
	Aliphatic GS		28.52 ±4.96	25.06 ±2.04	21.41 ±2.86	21.47 ±0.75	15.26 ±3.78	18.53 ±2.64	0.00	0.00	24.60 ±2.86	11.82 ±1.21
	Indolic GS		11.24 ±0.84	6.26 ±0.17	10.84 ±0.47	2.88 ±0.13	14.41 ±4.49	6.10 ±0.57	4.71 ±1.19	3.56 ±0.31	11.90 ±2.07	4.96 ±0.30
	Total GS		37.27 ±6.53	31.42 ±1.93	31.95 ±2.72	24.35 ±0.71	32.24 ±3.94	24.50 ±3.03	4.71 ±1.19	3.56 ±0.31	37.59 ±3.90	16.88 ±1.03

Table 5.5: Induction of GS contents in different host plants varieties upon *P. rapae* and *P. brassicae* herbivory

Host plant varieties	GS compounds	Fold change in GS content as a result of *P. rapae* herbivory	Fold change in GS content as a result of *P. brassicae* herbivory
***Brassica oleracea* var. *gemmifera* DC. 'Igor'**	Total	1.55	No change
	Aliphatic	No change	No change
	Indolic	2.47	1.80
	4-Methoxy-3-Indolylmethyl-GS	4.54	2.84
	1-Methoxy-3-Indolylmethyl-GS	33.38	36.64
***Brassica oleracea* var. *sabellica* L. 'Halbhoher Grüner Krauser'**	Total	1.79	No change
	Aliphatic	No change	No change
	Indolic	7.69	3.76
	4-Methoxy-3-Indolylmethyl-GS	3.52	1.55
	1-Methoxy-3-Indolylmethyl-GS	21.69	8.24
***Brassica oleracea* var. *medullosa* Thell. 'Grüner Ring'**	Total	2.37	No change
	Aliphatic	1.85	No change
	Indolic	3.90	2.36
	4-Methoxy-3-Indolylmethyl-GS	9.86	5.47
	1-Methoxy-3-Indolylmethyl-GS	19.57	7.80
***Brassica rapa* ssp. *pekinensis* 'Kasumi F1'**	Total	3.12	No change
	Aliphatic	-	-
	Indolic	3.12	No change
	4-Methoxy-3-Indolylmethyl-GS	8.50	3.39
	1-Methoxy-3-Indolylmethyl-GS	3.01	No change
***Brassica oleracea* var. *sabauda* L. 'Vorbote/Hilmar'**	Total	2.06	2.23
	Aliphatic	1.51	2.08
	Indolic	3.38	2.40
	4-Methoxy-3-Indolylmethyl-GS	2.73	No change
	1-Methoxy-3-Indolylmethyl-GS	34.24	14.37

Table 5.6: Pearson product-moment correlations between weight gain of *P. rapae* larval weight gain and each GS constituent's levels.

S.No.	GS	Pearson correlation with weight gain of *P. rapae* larvae	Significance (2 tailed)
1.	2-Hydroxy-3-Butenyl-	0.101	0.849
2.	2-propenyl-	0.164	0.757
3.	4-Methylsulfinylbutyl-	0.568	0.239
4.	3-Butenyl-	0.356	0.488
5.	4-Hydroxy-3-Indolylmethyl-	0.058	0.913
6.	3-Indolylmethyl-	0.565	0.243
7.	4-Methoxy-3-Indolylmethyl-	0.204	0.699
8.	1-Methoxy-3-Indolylmethyl-	-0.015	0.978
9.	Total GS content	-0.148	0.779

Table 5.7: Pearson product-moment correlations between weight gain of *P. brassicae* larval weight gain and each GS constituent's levels.

S.No.	GS	Pearson correlation with weight gain of *P. brassicae* larvae	Significance (2 tailed)
1.	2-Hydroxy-3-Butenyl-	-0.1	0.85
2.	2-propenyl-	-0.324	0.531
3.	4-Methylsulfinylbutyl-	0.532	0.277
4.	3-Butenyl-	-0.397	0.436
5.	4-Hydroxy-3-Indolylmethyl-	-0.419	0.409
6.	3-Indolylmethyl-	0.499	0.313
7.	4-Methoxy-3-Indolylmethyl-	-0.066	0.901
8.	1-Methoxy-3-Indolylmethyl-	-0.153	0.772
9.	Total GS content	-0.229	0.662

5.5 Discussion

In this study, three major topics were discussed, (1) Host-plant suitability of five Brassica varieties against two specialists, *P. rapae* and *P. brassicae* was investigated, (2) induced changes of GS levels in *Brassica* plants upon caterpillar feeding was analyzed, and 3) the correlation of the changes in total GS amounts of various *Brassica* varieties and larval weight was calculated.

The insect performance of both the specialists, feeding on five host plant varieties was examined by considering the weight gain by larvae during the experiment. We have seen different host-plant suitability of tested crop plants only for *P. brassicae*, not for *P. rapae*. The change in larval weight of *P. brassicae* showed that Marrow-stem Kale was found to be the best host plant variety for *P. brassicae*. *P. rapae* feeding on Marrow-stem Kale also showed a slightly better weight gain than other host plant varieties. Hence, it could be concluded that for both the specialist, Marrow-stem Kale was found to be the most suitable host plant variety among the five varieties, used for the experiment. For *P. brassicae*, Kale could be served as a second option for the host plant. Whereas, feeding on Savoy cabbage and Brussels sprouts resulted in the least weight gain when they were forced to feed on these host plant varieties for one week. This result agrees with a previous study by Lal and Chandra in which host plant susceptibility was checked and it was observed that *P. brassicae* of Indian originate very little and had low percentages reaching maturity when they fed two cultivars of Savoy cabbage, grown in India. Many other factors, other than weight gain of insects upon feeding are also involved in selecting the resistant and susceptible host plants such as the size of the larvae at maturity and the percentage of larvae that developed into imagines (Chandra and Lal 1977). The amount and quality of food consumed by herbivores influence their fitness, growth rates, developmental duration and probabilities of survival, performance in mating success, as well as timing and extent of reproduction and disposal ability (David and Gardiner 1962; Slansky and Scriber 1985, Hasan and Ansari 2011). The insect-food relationship could be quantitatively studied in terms of consumption index, relative growth rate, approximate digestibility, and efficiency of conversion of indigested food (Ansari et al. 2012).

The second aim of the current study was to investigate the effect of specialist herbivory on GS levels of host plant varieties. For this study, we considered eight known GS of the Brassicaceae family. A voluminous literature was found for the presence of many GS with wide versatility of chemistry in this plant family. (Kjaer 1976; Feeny et al. 1970; Fahey et al. 2001; McDanell et al. 1988; Traka and Mithen 2009; Kastell et al. 2013; Cartea et al. 2008). All the *Brassica* varieties selected for our study had mainly 2-hydroxy-3-butenyl GS as a major component of aliphatic GS. The main compound of indolyl GS in all *Brassica* varieties studied was indol-3-ylmethyl GS.

Through our results, we did not detect aliphatic GS in Chinese cabbage probably due to experimental conditions. The dominant component was 1-methoxyindol-3-ylmethyl GS has already been found in another subspecies of *B. rapa* (Mucha-Pelzer et al. 2010). Some researchers found that 2-propenyl GS is the principal GS in Kale varieties of Spain (Cartea et al. 2008).

As a result of plant response to infestation, plants produce increased levels of GS and their breakdown products to repel herbivores. These volatiles attract predators and serve for communicating among generalists, specialists, and predators (Fürstenberg-Hägg et al. 2013; Textor and Gershenzon 2009). The present study also reveals the induction of total GS contents in higher amounts upon herbivory of both species of Pieris caterpillar, which can attract other specialists also to damage the crop. The increase in GS levels in all host plant varieties as a result of L4-L5 instar larvae herbivory was more pronounced than the L3-L4 instar herbivory. The smaller instar stage of larvae causes less damage to the plants and so induces less GS content production as compared to the herbivory of larger instar. This suggests that more larval feeding induces large amounts of GS production, which could attract more specialists such as *Psylliodes chrysocephala* (cabbage stem flea beetle) which only feed on GS-containing plants (Bartlet and Williams 1991). In the body of herbivores, either GS contents remain stored throughout their life cycle as in the case of the specialist *P. chrysocephala*, or the plant GS content is hydrolyzed by the insect-produced myrosinase, as in *P. striolata*, which selectively accumulate GS from their food plants and express their myrosinase. The major substrates of *P. striolata* that produced myrosinase were aliphatic GS, which were hydrolyzed with at least fourfold efficiency than aromatic and indolic GS (Beran et al. 2018; 2014).

In the present study, aliphatic as well as indolic GS contents were also analyzed separately. The *P. rapae* herbivory induces the higher levels of aliphatic GS only in Marrow-stem Kale while *P. brassicae* herbivory produces higher levels in Marrow-stem Kale as well as in Savoy cabbage. Our results suggest that specific herbivory has an impact on the elevation of specific GS production in the host plant variety of Savoy cabbage. It was observed that *Brassica* varieties other than Marrow-stem Kale and Savoy cabbage showed no changes or a slight decrease in aliphatic GS content upon damage by herbivores. Likewise, no changes in aliphatic GS were observed upon feeding of *Phaedon cochleariae* (Fabricius) on one variety of *B. rapa* L. (Rostas and Hilker 2002). For Arabidopsis thaliana ecotype Col, no changes in aliphatic GS were observed upon *P. rapae* herbivory, while indolyl GS levels were elevated (Mewis et al. 2006).

The current study also reveals the elevation in indolic GS levels upon larval herbivory. The quantitative analysis of indolyl GS compounds reveals that the herbivory of both specialists can cause multiple-fold induction of indolyl GS. Our results showed similarities to previous work of Agrawal, showing greater accumulation of indolyl than aliphatic GS in *Arabidopsis lyrate* L. and *Brassica oleracea* L. upon damage by *P. rapae* (Agarwal and Kurashige 2003). The marked differences in plant responses to various insect herbivores may arise from differences in the biochemical composition of insect saliva (Alborn et al. 2003). Plants are also able to recognize compounds in insect oral secretions, which elicit more intense volatile responses than mechanical damage alone (De Moraes et al. 2001; Turlings et al. 1993). The presence of an elicitor like Volicitin was first identified in *Spodoptera exigua* (beet armyworm) oral secretions and is also found in larvae of *P. brassicae*, which can trigger the production of volatiles in the herbivore-injured plants (Turlings et al. 2000). Among all four indolyl, GS found in all host plant varieties, 3-indolylmethyl was found in major quantity. Upon feeding damages, the other two indolyl GS, that is, 4-methoxy-indolyl-3-methyl GS and 1-methoxy-indolyl-3-methyl GS were induced to several folds increase. With *P. rapae*, 1-methoxy-indolyl-3-methyl GS was increased about 30-fold in Savoy cabbage and Brussels sprouts and about 20-fold in Marrow-stem Kale and Kale. With *P. brassicae*, 1-methoxy-indolyl-3-methyl GS was increased about 30-fold in Brussels sprouts while in Savoy cabbage, Kale, and Marrow-stem Kale about ten-fold increase was observed. In Chinese cabbage, both specialists induced more the induction of 4-methoxy-indolyl-3-methyl GS than 1-methoxy-indolyl-3-methyl GS as compared to the controls. Likewise, *P. striolata* herbivory triggered indolyl-3-methyl and 1-methoxy-indolyl-3-methyl GS to several-fold in *Brassica* varieties (Beran 2011). And *P. Cruciferae* feeding on *B. napus* cotyledons was shown to induce indol-3-yl methyl GS 3.5-fold higher as compared to control (Bodnaryk 1992).

Summarizing both lepidopteran specialists feeding results, we can conclude that feeding of *P. rapae*, as well as *P. brassicae*, elicit the induction of higher indolyl GS levels while an increase in the aliphatic GS levels was variety dependent. This fact might indicate the stronger responsiveness of the indolyl GS biosynthesis pathway upon herbivory of both specialists (Mewis et al. 2006).

Agarwal described that the host plant suitability for specialist herbivores is also influenced by different types of GS as well as their concentrations (Agarwal and Kurashige 2003). Rohr found out that methylsufinyl GS containing *A. thaliana* ecotypes were more resistant to *Spodoptera exigua* and *P. brassicae* than the lines containing hydroxypropyl GS as main compounds. Our study reveals that the GS content and their presence in different concentrations in various varieties did not have any effect upon *P. rapae* as well as *P. brassicae* larvae. So, we have also calculated the correlation between the host plant suitability and induction GS contents upon specialists' herbivory. We observed the highest weight gain of specialist larvae and highest induction in the GS levels of the same host plant variety, but the correlation was found to be very slight and not significant. Our result is in agreement with the study of Santolamazza-Carbone. They found out that *P. rapae* and *Phyllotreta Cruciferae* (Goeze) were not affected by different GS content of different genotypes of a local variety of Kale from northwest Spain (Santolamazza-Carbone et al. 2014).

6 General Discussion

The central hypothesis of this thesis was to study the pheromone communication in *Pieris* species and to investigate the host plant suitability, which is interesting for integrated pest management. We studied the similarities and differences in the wing pheromones of two different species, *P. rapae*, and *P. brassicae*. For IPM strategies like mass trapping and mating disruption, the pheromones produced by female moths and male butterflies have been determined for several decades (Gaston et al. 1967; Shorey 1966; Grula et al. 1980; El - Sayed et al. 2003). Therefore, in this study, we wanted to explore the constitution of the pheromone blend produced by *P. rapae* and *P. brassicae* and four populations of different origins of *P. rapae*.

Identification of pheromone lures of P. rapae and P. brassicae - We have identified three main aphrodisiac compounds: ferrulactone, hexahydrofarnesylacetone (HHA), and E-phytol, found in the wings of male *P. rapae*. In contrast, another cyclic lactone named brassicalactone was produced in the wings of *P. brassicae*. The other two compounds HHA, and E-phytol, were also found in the wings of male *P. brassicae*. These all compounds showed an electroantennographic response in the present study, as discussed by Yildizhan (2009).

Ferrulactone in *P. rapae* and brassicalactone in *P. brassicae* were found to be male-specific wing components. Ferrulactone was found to be a major component of the aggregation pheromones of rust-red grain beetle, *Cryptolestes ferrugineus*, which is used to control the number of mature beetles in storehouses (Lindgren et al. 1985; Loschiavo et al. 1986). brassicalactone is also a cyclic ester like ferrulactone with an additional isoprene unit. Another component, E-phytol, is diterpene alcohol produced by only males of *P. rapae*. While, in *P. brassicae*, both sexes produce E-phytol. The males produce ten folds more than the females. Phytol is also found in plants as a secondary metabolite that is used to prepare chlorophyll a (Willstätter and Stoll 1913). E-phytol was also found in male wings extract of *Ephestia elutella*. A specific ratio of phytol with a γ-lactone elicited a female response equivalent to that produced by the unfractionated wing extract (Phelan et al. 1986).

The third compound, hexahydrofarnesylacetone (6, 10, 14-trimethylpentadecan-2-one), is produced by males and females in *P. rapae* and *P. brassicae*. It is a part of the male Danaine butterfly's pheromone system, *Idea leuconoe* (Schulz et al. 1993; Schulz and Nishida 1996). It was present in the perfume blends of fifteen species of orchid bees and a major component of nine of them (Zimmerman et al. 2009). During a detailed study of sex pheromones in male butterflies, *Bicyclus anynana*, with those of female moths, was found to be a part of male-specific pheromones (Nieberding et al. 2008). Hexahydrofarnesylacetone was also present in the hair pencil extracts of

Tirathaba mundella Walker (male oil palm bunch moth), and its antenatal response from females is also known (Sasaerila et al. 2003). It was also found in minor quantities in the pheromone volatiles of males of the *Galleria mellonella* (great wax moth) from 6 different regions of Russian regions (Lebedeva et al. 2002).

An aphrodisiac pheromone could also signal male quality. In some species, for example, in the genus *Nasonia*, the male aphrodisiac pheromone lacks species specificity, as heterospecific courting males may be accepted as a mate by a female. *Nasonia* species also possess remarkably similar sex pheromones, which leads to interspecific courtship (Buellesbach et al. 2014).

In other genera, such as *Leptopilina*, mate recognition is species-specific (Weiss et al. 2013; 2015) which usually prevents interspecific courtship. Isidoro proposed that a male antennal aphrodisiac pheromone also exists in *L. boulardi*. Their work demonstrated that antennal contact between males and females during courtship is required to elicit receptiveness in females. Additionally, they described glands and gland openings in the third and fourth male antennomeres. These antennomeres are brought into contact with the distal part of the female antennae during courtship (Isidoro et al. 1999). Thus, it is assumed that a chemical substance, an aphrodisiac pheromone, is transferred from the male antennae onto the female antennae to elicit female receptiveness. The species-specific investigations of the male courtship signals in *L. heterotoma*, *L. boulardi*, and *L. victoriae* have been done, and the antennal compounds of the males were analyzed (Weiss et al. 2015).

This species-specific difference between *P. rapae* and *P. brassicae* has already been identified from the Netherlands' populations (Yildizhan et al. 2009). The species-specific differences between male wings compounds have been identified for *Colias eurytheme* and *Colias philodice*. They are myristate, palmitate, and stearate in the wings of *C. philodice*, while 13-methyl heptacosane was present in *C. eurytheme* (Grula et al. 1980; Taylor 1973). The differences among the three species, *P. napi*, *P. rapae*, and *P. brassicae*, in transferring the anti-aphrodisiac pheromone have also been studied. The use of these chemicals curtails the courtship and decreases the likelihood of female re-mating. Methyl salicylate in *P. napi*, methyl sulfide, and indole in *P. rapae* while benzyl cyanide for *P. brassicae* (Andersson et al. 2003).

In integrated pest management, Methods used to concentrate the insects are pheromone baits, e.g., sex pheromones of lepidopteran females have been used to trap adult males. Both the moths, *Clepsis spectra* (Treitschke) and *Adoxophyes orana* (Fischer von Roslerstamm), with similar flight activity and response periods, produce the same two components of sex pheromone. These are

(Z)-9- and (Z)-11-tetradecenylacetate, but the males respond to different critical ratios (Minks et al. 1973). The four components of the female sex pheromone of *Archips argyrospilus* (Walker) and *Archips mortuanus* (Kearfoot) are (Z)- and (E)- 11 tetradecenyl acetates, (Z)-9- tetradecenyl acetate, and dodecanyl acetate. The males of *A. argyrospilus* are attracted to a specific blend ratio, that is 60:40:4:200. In contrast, the males of *A. mortuanus* were attracted to 90:10:1:200. Shifts in these optimal ratios cause a drastic reduction in trap catches (Carde et al. 1977; Carde and Millar 2009). Hence getting knowledge about a perfect pheromone lure for a pest is important to carry out a quantification study of important pheromone components.

According to our research, the pheromone blend ratio of *P. brassicae* was found to be 1:23:233:2 of brassicalactone, hexahydrofarnesylacetone, E-phytol, and isophytol. This quantitative difference helps in the process of identifying a suitable partner of the same species and population for courtship. The main components of volatiles collected from calling males of great wax moth were nonanal and undecanal, and their ratio found in population from Canada were 1:3 (Romel et al. 1992), while in the population from the USA, it was 7:3 (Leyrer and Monroe 1973). Furthermore, the Russian population consists of also same components with different ratios for all six regions in Russia (Lebedeva et al. 2002). The response of females to the synthetic bait in laboratory tests was not as high as their response to living males, and in field tests, the mixture was also inactive. (Flint and Merkle 1983; Finn and Payne 1977).

Geographic variations in chemical communication of Asian and European populations of P. rapae - The quantitative variations in aphrodisiac compounds among the diverse populations of *P. rapae* from various geographic regions were also among the objectives of this study. In our study, the ratio of pheromone blend of male *P. rapae*, the German population was found out to be 1.5:1.9:1 for ferrulactone, hexahydrofarnesylacetone, and phytol. For the same compounds, the blend for the Netherlands population was 1.0:3.4:0. For Taiwan's population, it was 1.0:4.5:8.0, and for the Vietnam population, the blend ratio was 1.0:1.3:4.3. Hence, we can conclude that these ratios for all population blends are different.

As in the past few years, case studies have been done in some Asian and African countries to check the influence of geographic regions on populations. The synthetic blend of pheromones was checked for *Maruca vitrata*. However, it did not attract males in field trapping experiments in Taiwan. The same lures did not attract males in Thailand and Vietnam either (Srinivasan et al. 2015). This blend of pheromones was the most attractive lure in field studies in Benin and attracted a total of 33.1 males per trap over eight weeks (Downham et al. 2004). These findings indicate geographic variation in the sex pheromone blend between Asian and West African *M. vitrata* populations.

Likewise, the geographical variations in the pheromone system of *Hemileuca eglanterina* (Boisduval) have been explored. There are two distinct types of pheromone systems. Male moths from the population in the San Gabriel mountains of southern California are attracted optimally to a blend of E10, Z12-hexadeca-10,12-dien-l-yl acetate (acetate); E10, Z12-hexadeca-10,12-dien-l-ol (alcohol); and E10, Z12-hexadeca-10,12-dienal (aldehyde), whereas males from the population at Robinson Summit, northwest of Ely, Nevada, are attracted to the two-component blend of alcohol and aldehyde (McElfresh and Millar 2001).

Behavioral response of large cabbage white in response to some pheromone components - To use the determined ratios of male pheromone compounds against female Pierids in IPM strategies, these compounds must first be checked the female antennae response to them. Then these compounds have to be checked through some bioassays. In our study, electroantennographic responses of female butterflies were positive for all these compounds singly. A behavioral bioassay was then carried out using female *P. brassicae* against some E-phytol ratios with a mixture of minor compound combinations, hexahydrofarnesylacetone, and isophytol. The bioassay results reveal that the females are not attracted to single pure compounds but the mixtures. Sometimes, they did not even take a flight for the single compounds, and sometimes they took a unidirectional flight overall in the wind tunnel. As in the males' wings extracts, we have found E-phytol as a major component quantitatively, so we preferred to use different proportions of E-phytol for the bioassay. Therefore, for the mixtures, we have used E-phytol as a main component while hexahydrofarnesylacetone and isophytol as minor components. For the mixtures, the females were attracted to the ratios with 1:1:30 and 1:1:70 only. A few of them were flattering around and resting in the very close vicinity with the dummy male with 1:1:70 mixture. Likewise, they were reacting to the wings extract. Hence, the females seem to reject the single components produced by males' wings. However, it can be concluded that the females accept the males during courtship depending upon the whole bouquet or some specific ratio of some components from that bouquet they release. As showed by Yildizhan in a mating-competitive bioassay, which proves about 77% of mated males had received the specific ratios of pheromone content in which they used brassicalactone, hexahydrofarnesylacetone, and E-phytol.

Female receptivity is usually signaled by a combination of visual and olfactory signals, visual stimuli being more important at a distance and olfactory stimuli being more important close up (Silberglied 1984). The male butterflies use aphrodisiac pheromones to attract the females to bring them finally towards courtship. So, these chemicals worked at a distance to attract the conspecifics and mate recognition as it was showed in genus *Pieris* when *P. melete* males attempted more frequently to copulate with its conspecific females than with the *P. rapae* (Ohguchi and Hidaka 1988).

In the Pieridae, both virgin and mated females adopt a characteristic posture when courted, spreading the wings horizontally and lifting the abdomen vertically, a posture well designed for emitting chemical stimuli close to the location of the hovering courting male (Wiklund and Forsberg 1985; Forsberg and Wiklund 1989). We have found the same response of female butterflies with male pheromone extract and one of the mixture ratios used for the bioassay.

In conclusion, our study supports the variations in the sexual communication used to attract opposite sexes between species and various Asian and European populations. In the future, it would be suggested to find out some specific pheromone lures for specific populations through field experiments.

Host plants' suitability and their response to herbivory in terms of GS - Another prospect of studying chemical ecology is to explore the plant-insect relationship, which is important to develop pest control strategies. The insect pest management strategies include pest monitoring, which allows accurate timing of pesticide applications; combined use of semiochemicals, host-plant resistance, trap crops to manipulate pest behavior, employing biological control approaches, or selective insecticides to reduce pest populations and developing insect-resistant crops (Pickett et al. 1997). These all techniques can be developed for a *brassica* pest only when the host selection behavior of herbivores on Brassica host plants and the insect infestation responses of *brassica* crops are known (Ahuja et al. 2010). In the current study, we have seen different host-plant suitability of tested crop plants only for *P. brassicae*, not *P. rapae*. The change in larval weight of *P. brassicae* showed that Marrow-stem Kale was found to be the best host plant variety for *P. brassicae*. *P. rapae* feeding on Marrow-stem Kale also showed a slightly better weight gain than other host plant varieties. Hence, it could be concluded that for both the specialist, Marrow-stem Kale was found to be the most suitable host plant variety among the five varieties used for the experiment. For *P. brassicae*, Kale could be served as a second option for the host plant. Whereas feeding on Savoy cabbage and Brussels sprouts resulted in minor weight gain, they were forced to feed on these host plant varieties for one week.

Our study leads to exploring the variations in Glucosinolate (GS) content upon herbivory of two specialist *P. rapae* and *P. brassicae* larvae. The profiling of GS contents was carried out in different brassica cultivars. We have found four aliphatic and four indolic GS altogether with a great variance in their quantitative levels. Aliphatic GS includes 2-hydroxy-3-butenyl (progoitrin), 2-propenyl (sinigrin), 4-Methylsulfinylbutyl (glucoraphanin) and 3-Butenyl (gluconapin) GS. 2-propenyl GS (Sinigrin) is the principal GS found in Kale varieties of Spain (Cartea et al. 2008). According to research on Arabidopsis thaliana L. (Brassicaceae), it produces three GS, that is methylsulfinyl, 3-hydroxypropyl, and 2-propenyl GS (Sinigrin) (Rohr et al. 2009). Glucoraphanin and 4-hydroxyglucobrassicin were quantified in seeds of 33 commercially available cultivars of broccoli, four cultivars of Turnip cabbage, Radish, Cauliflower, Brussels sprouts, Kale, and Cabbage, and two cultivars of Raab (West et al. 2004).

While analyzing different *Brassica* varieties, we have found four different indolyl GS. One was indol-3-ylmethyl (glucobrassicin), and three others were the derivatives of glucobrassicin, 4-Hydroxyindol-3-ylmethyl (4-hydroxyglucobrassicin), 4-methoxyindol-3-ylmethyl (4-methoxyglucobrassicin), and 1-methoxyindol-3-ylmethyl (neoglucobrassicin) GS. Glucobrassicin on the surface of cabbage (*Brassica oleracea*) leaves potent stimulant-inducing oviposition by the cabbage butterflies *P. brassicae* (van Loon et al. 1992) and *P. rapae* (Renwick et al. 1992).

Previous studies have confirmed that GS plays a vital role in host selection by several insect species. For example, oviposition and larval feeding by *P. napi macdun- noughii* correlate with GS profiles of species in a natural community (Rodman and Chew 1980). Different GS may play various roles in host selection by a particular insect species. It was also found that Sinigrin could trigger the host selection behavior of *P. rapae*, but glucobrassicin produced a much stronger effect (Traynier and Truscott 1991). Plant responses upon pest infestation were also studied in detail. According to a study, plants release GS to repel herbivores, attract predators or communication between leaves or plants and induce defense responses (Arimura et al. 2004). GS act as repellents of generalists and attractants of specialists such as *Psylliodes chrysocephala* (cabbage stem flea beetle) only feed on GSL containing plants (Turlings et al. 1993). In this study, the quantitative analysis of GS components reveals a several-fold increase in aliphatic GS. The indolic GS levels were also increased significantly upon P. rapae larval infestation. The increase in aliphatic GS was found to be more than indolic GS levels. Whereas with *P. brassicae* feeding, we did not find a significant difference in aliphatic GS amounts for all the varieties except for Savoy cabbage, in which aliphatic GS levels increased two-fold. For Arabidopsis thaliana ecotype col, no changes in aliphatic GS and only a slight increase in indolyl GS upon *P. rapae* herbivory was documented (Mewis et al. 2006). We found a several folds increase in indolic GS amounts in all host plant varieties used for this study. Our results showed resemblance with the previous work of Agrawal and Kurashige, showing a more significant accumulation of indolyl than aliphatic GS in *Arabidopsis lyrata* L. and *Brassica oleraceae* L. upon damage by *P. rapae* (Agarwal and Kurashige 2003).

Comparing both specialist feeding results, we can conclude that feeding of *P. rapae* did not affect the aliphatic GS levels except only one variety, and the feeding of *P. brassicae* also caused elevation only in selected cultivars. Both specialists elicited a several-fold increase in the indolyl GS levels in all *Brassica* varieties. This might indicate a more robust responsiveness of both specialists on the biosynthesis of indolyl GS (Mewis et al. 2006).

Agarwal described the host plant's suitability for specialist herbivores as influenced by different GS types and their concentrations (Agarwal et al. 2003). Santolamazza-Carbone found out that high content of Sinigrin decreased the abundance of *Mamestra brassicae* and *Plutella xylostella* while *P. rapae* and *Phyllotreta Cruciferae* were not affected by the different genotypes of a local variety of Kale from northwest Spain (Santolamazza-Carbone et al. 2014). Rohr found out that methyl sulfinyl GS containing ecotypes were more resistant to *Spodoptera exigua* and the *P. brassicae* than the lines containing hydroxypropyl GS as main compounds (Rohr et al. 2006). We also studied the correlation between specific GS constituents and larval weight gain of the specialist *Pieris* species. We did not find any significant correlation between the specific GS constituent and any of the specialist species. All the cultivars responded to *P. rapae* as well as *P. brassicae* strongly. The plants produced slightly more aliphatic GS when *P. rapae* than *P. brassicae* fed them. Considering the indolic GS levels, the plants reacted the same upon both specialists feeding. For both the species, indolic GS were increased to about two-folds in most cases. However, it can be concluded that all the host plant varieties respond to the different pest infestations differently. Both the species cause different changes in the GS contents upon feeding for the same time period.

In conclusion, the study of analyses of GS hydrolyzed products and behavioral activity of specialists to those candidate compounds open doors to future research.

7 References

Aftab MG, Ulrichs Ch, Schulz H, Gasch T, Mewis I (2016) Variations in the chemical profile of aphrodisiac pheromones in the wings of *Pieris rapae* populations of different geographic origin. 60. Pflanzenschutztagung Halle, Julius-Kühn-Archiv 454: 493.

Agerbirk N, De Vos M, Kim JH, Jander G (2009) Indole glucosinolate breakdown and its biological effects. Phytochem Rev 8: 101-120.

Agrawal AA, Kurashige NS (2003) A role for isothiocyanates in plant resistance against the specialist herbivore *Pieris rapae*. J Chem Ecol 29: 1403-1415.

Ahuja I, Rohloff J, Bones AM (2010) Defence mechanisms of Brassicaceae: implications for plant-insect interactions and potential for integrated pest management. A review. Agron Sustain Dev 30: 311-348.

Ahuja I, van Dam NM, Winge P, Trælnes M, Heydarova1 A, Rohloff J, Langaas M, Bones AM (2015) Plant defence responses in oilseed rape MINELESS plants after attack by the cabbage moth *Mamestra brassicae*. J Exp Bot 66(2): 579-592.

Alborn HT, Brennan MM, Tumlinson JH (2003) Differential activity and degradation of plant volatile elicitors in regurgitant of tobacco hornworm (*Manduca sexta*) larvae. J Chem Ecol 29: 1357-1372.

Andersson J 2000

Andersson J, Borg-Karlson AK, Vongvanich N, Wiklund C (2007) Male sex pheromone release and female mate choice in a butterfly. J Exp Biol 210: 964-970.

Andersson J, Borg-Karlson AK, Wiklund C (2003) Antiaphrodisiacs in Pierid butterflies: A theme with variation. J chem Ecol 29 (6): 1489-1499.

Ansari MS, Hasan F, Ahmad N (2012) Influence of various host plants on the consumption and utilization of food by *Pieris brassicae* (L.) Bull Entomol Res 102: 231-237.

Aono N, Natsumi K, Yukawa Y (1989) Mating suppression to the cherry tree borer, *Synanthedon hector* Butler, in Japanese apricot by its synthetic sex pheromone. Plant Prot 43: 329-332 (in Japanese).

Arimura G, Ozawa R, Kugimiya S, Takabayashi J, Bohlmann J (2004) Herbivore-induced defense response in a model legume. Two-spotted spider mites induce emission of (E)-β-ocimene transcript accumulation of (E)-β-ocimene synthase in *Lotus japonicus*. Plant Physiol 135: 1976-1983.

Atalay R; Hincal P (1992) Investigations on the species of Pieridae (Lepidoptera) which are harmful to plants belonging to the family Cruciferae with their importance in Izmir and its vicinity and the biology of the large white butterfly (*Pieris brassicae* L.) along with the factors affecting its population fluctuations. Doga Turk Tarim ve Ormancilik Dergisi 16(1): 271-286.

Avidov Z, Harpaz I (1969) Plant pests of Israel. Israel University Press.

Babushok VI, Linstrom PJ, Zenkevich IG (2011) Retention indices for frequently reported compounds of plant essential oils. J Phys Chem Ref Data 40(4) 043101-1-47.

Bartlet E, Williams IH (1991) Factors restricting the feeding of the cabbage stem flea beetle (*Psylliodes chrysocephala*). Entomol Exp Appl 60: 233-238.

Bellostas N, Sørensen AD, Sørensen JC, Sørensen H, Sørensen MD, Gupta SK, Kader JC (2007) Genetic variation and metabolism of glucosinolates. Adv Bot Res 45: 369-415.

Beran F (2011) Host preference and aggregation behavior of the striped flea beetle, *Phyllotreta striolata*. PhD Dissertation, Humboldt Universität zu Berlin, Berlin.

Beran F, Pauchet Y, Kunert G, Reichelt M, Wielsch N, Vogel H, Reinecke A, Svatov A, Mewis I, Schmidt D, Ramasamy S, Ulrichs C, Hansson BS, Gershenzon J, Heckel DG (2014) *Phyllotreta striolata* flea beetles use host plant defense compounds to create their own glucosinolate-myrosinase system. Proceedings of the National Academy of Sciences. 111(20): 7349-7354.

Beran F, Sporer T, Paetz C, Ahn S-J, Betzin F, Kunert G, Shekhov A, Vassão DG, Bartram S, Lorenz S, and Reichelt M (2018) One pathway is not enough: The cabbage stem flea beetle *Psylliodes chrysocephala* uses multiple strategies to overcome the glucosinolate-myrosinase defense in its host plants. Front Plant Sci 9: 1754.

Bergström G, Lundgren L (1973) Androconial secretion of three species of butterflies of the genus *Pieris* (Lep., Pieridae). Zoon Uppsala 1: 67-75.

Bhandari K, Sood P, Mehta P, Choudhary A, Prabhakar C (2009) Effect of botanical extracts on the biological activity of granulosis virus against *Pieris brassicae*. Phytoparasitica 37: 317-322.

Biever KD, Wilkinson JD (1978) A stress-induced granulosis virus of *Pieris rapae*. Environ Entomol 7: 572-573.

Bodnaryk RP (1992) Effects of wounding on glucosinolates in the cotyledons of oilseed rape and mustard. Phytochemistry 31: 2671-2677.

Boeckh J, Ernst KD (1987) Contribution of single unit analysis in insects to an understanding of olfactory function. J Comp Physiol 161: 549-565.

Bouwer MC, Slippers B, Degefu D, Wingfied MJ, Lawson S, Rohwer ER (2015) Identification of the sex pheromone of the tree infesting cossid moth *Coryphodema tristis* (Lepidoptera: Cossidae). PLoS ONE 10(3): e0118575.

Branca F, Lib G, Goyalc S, Quiros CF (2002) Survey of aliphatic glucosinolates in Sicilian wild and cultivated Brassicaceae. Phytochemistry 59: 717-724.

Buellesbach J, Greim C, Raychoudhury R, Schmitt T (2014) Asymmetric assortative mating behaviour reflects incomplete pre-zygotic isolation in the *Nasonia* species complex. Ethology 120(8) 834-843.

Burgess L, Wiens JE (1980) Dispensing allyl isothiocyanate as an attractant for trapping crucifer-feeding flea beetles. Canadian agriculture library Research Station, Agriculture Canada, Saskatoon, Saskatchewan S7N 0X2 (Canada).

CABI, Centre of agriculture and bioscience international; 2018.

Capinera JL (2008) Butterflies and moths. Encyclopedia of entomology, 4 (II ed) Springer 626-672.

Cardé RT, Cardé AM, Hill AS, Roelofs WL (1977) Sex pheromone specificity as a reproductive isolating mechanism among the sibling species *Archips argyrosphilus* and *A. mortuanus* and other sympatric Tortricine moths (Lepidoptera: Tortricidae). J Chem Ecol 3: 71-84.

Cardé RT, Millar JG (2009) Pheromones. Encyclopedia of Insects 766-772.

Cardé RT, Minks AK (1995) Control of moth pests by mating disruption: Successes and constraints. Annu Rev Entomol 40: 559-585.

Cartea ME, Velasco P, Obregón S, Padilla G, De Haro A (2008) Seasonal variation in glucosinolate content in *Brassica oleracea* crops grown in northwestern Spain. Phytochemistry 69: 403-410.

Chandra J, Lal OP (1977a) Development and survival of caterpillars of cabbage butterfly, *P. brassicae* L. on some varieties of cabbage. Indian J Ent 38: 187-188.

Chandra J, Lal OP (1977b) Mating behaviour of cabbage butterfly, *Pieris brassicae* Linn. Indian J Ent 38: 197-198.

Chen X, Nakamuta K, Nakanishi T, Nakashima T, Tokoro M, Mochizuki F, Fukumoto T (2006) Female sex pheromone of a carpenter moth, *Cossus insularis* (Lepidoptera: Cossidae). J Chem Ecol 32: 669-679.

Cork A (1994) Identification of electrophysiologically-active compoundsfor New World screwworm, Cochliomyia hominivorax, in larval woundfluid. Med. Vet. Entomol., 8, 151–159.

Cole RA (1976) Isothiocyanates, nitriles, and thiocyanates as products of autolysis of glucosinolates in Cruciferae. Phytochemistry 15: 759-762.

Cole RA (1997) The relative importance of glucosinolates and amino acids to the development of the aphid pests *Brevicoryne brassicae* and *Myzus persicae* on wild and cultivated *Brassica* species. Entomol Exp Appl 85: 121-133.

Conner WE, Eisner T, Vander Meer RK, Guerrero A, Chiringelli D, Meinwald J (1980) Sex attractant of an arctiid moth *Utethesia ornatrix*: a pulsed chemical signal. Behav Ecol Sociobiol 7: 55-63.

Cossé A, Todd J, Millar J, Martinez L, Baker T (1995) Electroantennographic and coupled gas chromatographic-electroantennographic responses of the Mediterranean fruit fly, *Ceratitis capitata*, to male-produced volatiles and mango odor. J Chem Ecol 21: 1823-1836.

Cranshaw WS, Default RJ (1985) Effect of artificial defoliation on broccoli yield. Great lakes Entomol 18: 55-57.

David WAL, Gardiner BOC (1961) The mating behaviour of *Pieris brassicae* (L.) in a laboratory culture. Bull Ent Res 52: 263-280.

David WAL, Gardiner BOC (1962) Observation on the larva and pupa of *Pieris brassicae* (L.) in the laboratory culture. Bull Ent Res 53: 414-436.

Davies CR, Gilbert N (1985) A comparative study of the egg-laying behaviour and larval development of *Pieris rapae* L. and *Pieris brassicae* L. on the same host plants. Oecologia (Berlin) 67: 278-281.

Daxenbichler ME, Van Etten CH, Spencer GF (1977) Glucosinolates and derived products in cruciferous vegetables. Identification of organic nitriles from cabbage. J Agric Food Chem 25: 121-124.

De Moraes CM, Mescher MC, Tumlinson JH (2001) Caterpillar-induced nocturnal plant volatiles repel conspecific females. Nature 410: 577-580.

Dhaliwal GS, Dhawan AK, Singh R (2007) Biodiversity and ecological agriculture: Issues and perspectives. Indian J Ecol 34(2): 100-109.

Dicke M, Vet LEM (1999) Plant-carnivore interactions: Evolutionary and ecological consequences for plant, herbivore and carnivore. In Olff H, Brown VK, and Drent RH (ed). Herbivores between plants and predators. Blackwell Science. 483-520.

Downham MCA, Tamò M, Hall DR, Datinon B, Adetonah S, Farman DI (2004) Developing pheromone traps and lures for *Maruca vitrata* in Benin, West Africa. Entomol Experi Appl 110: 151-158.

Drijfhout F (2017) Chemical ecology. In: Encyclopedia of life sciences (ELS). Wiley, Chichester. https://doi.org/10.1002/9780470015902.a0003265.pub3.

Dusaussoy G, Delplanque A (1964) L elevage de *Pieris brassicae* L. en toutes saisons: accouplement et ponte en conditions artificielles. Rev path veg ent agric Fr 43: 119-134.

Eliot N (1944) Larval anthropomorphosis. Entomologist 77: 22-27.

El-Sayed AM, Delisle J, De Lury N, Gut LJ, Judd GJR, Legrand S, Reissig WH, Roelofs WL, Unelius CR, Trimble RM (2003) Geographic variation in pheromone chemistry, antennal electrophysiology, and pheromone-mediated trap catch of North American populations of the oblique-banded leafroller. Environ Entomol 32 (3): 470-476.

European food safety authority (2008) Glucosinolates as undesirable substances in animal feed scientific panel on contaminants in the food chain. The EFSA Journal 590: 11-76.

Escofet M, Boada J, Melo JC, Casals C, Barrios G, Palau G, Cavallé C, Pelaez M, Mateu J, Aymami A, Nubiola N, Ribé E, Ibáñez R, Santiveri C, Sans J, Pallarés J, Agustí JA (2005) Sexual disruption to control leopard moth (*Zeuzera pyrina*) in hazelnut. Acta Hortic 686: 421-425.

Fabre JH (1912) The life of the caterpillar. Hodder and Stoughton.

Fahey JW, Zalcmann AT, Talalay P (2001) The chemical diversity and distribution of glucosinolates and isothiocyanates among plants. Phytochemistry 56: 5-51.

Feeny P, Paauwe KL, Demong NJ (1970) Flea beetles and mustard oils- host plant specificity of *Phyllotreta Cruciferae* and *Phyllotreta Striolata* adults (Coleoptera: Chrysomelidae). Ann Entomol Soc Am 63(3): 832-841.

Feltwell J (1982) Large white butterfly: The biology, biochemistry and physiology of *Pieris brassicae* (Linnaeus). Dr W. Junk Publishers the Hague, Boston, London.

Fenwick GR, Heaney RK, Mullin WJ (1983) Glucosinolates and their breakdown products in food and food plants. Crit Rev Food Sci Nutr 18(2): 123-201.

Ferro DN (1993) Integrated pest management in vegetables in Massachusetts. Successful implementation of integrated pest management for agricultural crops by Leslie AR and Cuperus GW (ed). 95-105.

Finn WE, Payne TL (1977) Attraction of greater wax moth females to male produced pheromones. Southwest Entomol 2: 62-65.

Flint HM, Merkle JR (1983) Mating behavior, sex pheromone responses and radiation sterilization of the greater wax moth (Lepidoptera: Pyralidae). J Econ Entomol 76: 467-472.

Ford HD (1920) Entomologist 53: 139 (scarcity).

Ford HD (1945) Butterflies. Collins, London.

Forsberg J, Wiklund C (1989) Mating in the afternoon: Time-saving in courtship and remating by females of a polyandrous butterfly, *Pieris napi*. Behav Ecol Sociobiol 25: 349-356.

Fukano Y, Satoh T, Hirota T, Nishide Y, Obara Y (2012) Geographic expansion of the cabbage butterfly (*Pieris rapae*) and the evolution of highly UV-reflecting females. Journal of Insect Science 19: 239-246.

Fürstenberg-Hägg J, Zagrobelny M, Bak S (2013) Plant Defense against insect herbivore species. Int J Mol Sci 14: 10242-10297.

Gaston L, Shorey H, Saario C (1967) Insect population control by the use of sex pheromones to inhibit orientation between the sexes. Nature 213: 1155.

Gardiner BOC (1978) Instar number and pupal colouration in Palestinian *Pieris brassicae*. Proc & Trans Brit ent nat Hist Soc 11: 21-23.

Gols R, Bukovinszky T, Van Dam N, Dicke M, Bullock J, Harvey J (2008) Performance of generalist and specialist herbivores and their endoparasitoids differs on cultivated and wild *Brassica* Populations. J Chem Ecol 34: 132-143.

Gols R, Harvey J (2009) Plant-mediated effects in the Brassicaceae on the performance and behaviour of parasitoids, Phytochem Rev 8: 187-206.

Grula JW, McChesney JD, Taylor OR (1980) Aphrodisiac pheromones of the Sulphur butterflies *Colias eurytheme* and *Colias philodice* (Lepidoptera: Pieridae) J Chem Ecol 6(1): 241-256.

Halkier BA, Gershenzon J (2006) Biology and Biochemistry of glucosinolates. Annu Rev Plant Biol 57: 303-333.

Hanschen FS, Schreiner M (2017) Isothiocyanates, nitriles, and epithionitriles from glucosinolates are affected by genotype and developmental stage in *Brassica oleracea* Varieties. Front plant sci 8: 1095.

Harcourt DG (1966) Major factors in survival of immature stages of *Pieris rapae* (L.). Can Entomol 98: 653-662.

Harcourt DG, Backs RH, Cass LM (1955) Abundance and relative importance of caterpillars attacking cabbage in eastern Ontario. Can Entomol 87: 400-406.

Harvey JA, Jervis MA, Gols R, Jiang N, Vet LEM (1999) Development of the parasitoid, *Cotesia rubecula* (Hymenoptera: Braconidae) in *Pieris rapae* and *Pieris brassicae* (Lepidoptera: Pieridae): evidence for host regulation. J Insect Physiol 45: 173-182.

Harvey JA, Biere A, Fortuna T (2010) Ecological fits, misfits and lotteries involving insect herbivores on the invasive plant, *Bunias orientalis*. Biol Invasions 12: 3045-3059.

Hasan F, Ansari MS (2010) Effect of different Cole crops on the biological parameters of *Pieris brassicae* L. (Lepidoptera: Pieridae) under laboratory conditions. J Crop Sci Biotechnol 13(3): 195-202.

Hasan F, Ansari MS (2011) Population growth of *Pieris brassicae* L. (Lepidoptera: Pieridae) on different Cole crops under laboratory conditions. J Pest Sci 84(2): 179-186.

Hegazi E, Khafagi W, Konstantopoulou M, Schlyter F, Raptopoulos D, Shweil S, El-Rahman S, Atwa A, Ali S, Tawfik H (2010). Suppression of Leopard moth (Lepidoptera: Cossidae) populations in Olive trees in Egypt through mating disruption. Journal Econ Entomol 103: 1621-1627.

Hely PC; Pasfield G; Gellatley JG, 1982. Insect Pests of Fruit and Vegetables in NSW. Sydney, New South Wales, Australia: Department of Agriculture, New South Wales, viii + 312.

Heppner JB (2008) Butterflies and Moths (Lepidoptera). In: Capinera J.L. (ed) Encyclopedia of Entomology. Springer, Dordrecht. https://doi.org/10.1007/978-1-4020-6359-6_498.

Hertz M (1927) Huomioita petokuoriaisten olinpaikoista. Luonnon Ystava 31: 218-22.

Hewitt CG (1917) Insect behavior as a factor in applied entomology. J econ ent concord N H 10: 1.

Hildebrand JG, Shepherd GM (1997) Mechanisms of olfactory discrimination: converging evidence for common principles across phyla. Annu Rev Neurosci 20: 595-631.

Hill DS (1987) Agricultural insect pests of temperate regions and their control. Cambridge, UK: Cambridge University Press.

Hölldobler B, Wilson EO (1990) The Ants. Springer Verlag, Berlin.

Honda K, Kawatoko M (1982) Exocrine substances of the white cabbage butterfly, *Pieris rapae* crucivora. Appl Entomol Zool 17: 325-331.

Honda K, Omura H, Hayashi N (1998) Identification of floral volatiles from *Ligustrum japonicum* that stimulate flower-visiting by cabbage butterfly, *Pieris rapae*. J Chem Ecol 24(12): 2167-2180.

Honda K, Omura H, Hayashi N (1999) Identification of floral volatiles from *Ligustrum japonicum* that stimulate flower-visiting by cabbage butterfly, *Pieris rapae*. J Chem Ecol 24: 2167-2180.

Hopkins RJ, Griffiths DW, Birch ANE, McKinlay RG (1998) Influence of increasing herbivore pressure on modification of glucosinolate content of swedes (*Brassica napus* spp. *rapifera*). J Chem Ecol 24: 2003-2019.

Hopkins RJ, Van Dam NM, Van Loon JJA (2009) Role of glucosinolates in insect-plant relationships and multitrophic interactions. Annu Rev Entomol 54: 57-83.

Hoshi H, Takabe M, Nakamuta K (2016) Mating Disruption of a Carpenter Moth, *Cossus insularis* (Lepidoptera: Cossidae) in Apple Orchards with Synthetic Sex Pheromone, and Registration of the Pheromone as an Agrochemical. J Chem Ecol 42: 606-611.

Isidoro N, Bin F, Romani R, Pujade-Villar J, Ros-Farre P (1999) Diversity and function of male antennal glands in Cynipoidea (Hymenoptera). Zoologica Scripta 28(1-2): 165-174.

Ives PM (1978) How discriminating are cabbage butterflies? Aust J Ecol 3: 365-397.

Jackson WH (1890) IV. Studies in the morphology of the lepidoptera. Part I. Trans Linn Soc Lond 4: 143.176.

Jankowska B (2006) The occurrence of some lepidoptera pests on different cabbage vegetables. J Plant Prot Res 46(2): 181-190.

Jones RE (1977) Movement patterns and egg distribution in cabbage butterflies. J Anita Ecol 46: 195-212.

Kanzaki R, Sugi N, Shibuya T (1992) Self-generated zigzag turning of Bombyx mori males during pheromone-mediated upwind walking. Zool Sci 9(3): 515-527.

Karlson P, Betenandt A (1959) Pheromones (ectohormones) in insects. Annu Rev Entomol 4: 39-58.

Karlson P, Lüscher M (1959) Pheromones: A new term for a class of biologically active substances. Nature 183: 55-5.

Karlson P, Schneider D (1973) Sexual pheromones of Lepidoptera as model systems for chemical communication. Naturwissenschaften 60: 113-121.

Karowe DN, Schoonhoven LM (1992) Interactions among three trophic levels: the influence of host plant on performance of *Pieris brassicae* and its parasitoid, *Cotesia glomerata*. Entomol Exp Appl 62(3): 241-251.

Kastell A, Smethanska I, Ulrichs C, Cai Z, Mewis I (2013) Effects of phytohormones and jasmonic acid on glucosinolate content in hairy root cultures of *Sinapis alba* and *Brassica rapa*. Appl Biochem Biotechnol 169: 624-635.

Kelly PJ, Bones A, Rossiter JT (1998) Sub-cellular immunolocalization of the glucosinolate sinigrin in seedlings of *Brassica juncea*. Planta 206: 370-377.

Khan HH, Kumar A, Naz H (2017) Population dynamic of cabbage butterfly (*Pieris brassicae* L.) in district Sultanpur (U.P.) A review. J entomol zool stud 5(6): 1938-1940.

Kim JH, Jander G (2007) *Myzus persicae* (green peach aphid) feeding on *Arabidopsis* induces the formation of a deterrent indole glucosinolate. Plant J 49: 1008-1019.

Kirby RD, Slosser JE (1984) Composite economic threshold for three lepidopterous pests of cabbage. J Econ Entomol 77(3): 725-733.

Kjaer A (1976) Glucosinolates in the Cruciferae. In: Vaughan JG, MacLeod AJ, Jones BMG (ed), The biology and chemistry of the Cruciferae. Academic Press, London 207-219.

Klein HZ (1932) Studien zur Ökologie und Epidemiologie der Kohlweißlinge. I. Der Einfluss der Temperatur und Luftfeuchtigkeit auf Entwicklung und Mortalität von *Pieris brassicae* L. Z angew ent 19: 395-448.

Koritsas VM, Lewis JA, Fenwick GR (1989) Accumulation of indole glucosinolates in *Psylliodes chrysocephala* L. -infested, or -damaged tissues of oilseed rape (*Brassica napus* L.) Experientia 45(5): 493-495.

Koritsas VM, Lewis JA, Fenwick GR (1991) Glucosinolate responses of oilseed rape, mustard and kale to mechanical wounding and infestation by cabbage stem flea beetle (*Psylliodes chrysocephala*). Ann Appl Biol 118: 209-221.

Kovats E (1958) Gas-chromatographische Charakterisierung organischer Verbindungen. Teil 1: Retentionsindices aliphatischer Halogenide, Alkohole, Aldehyde und Ketone. Helv Chim Acta 41(7): 1915-1932.

Lal MN, Bhajan R (2004) Cabbage butterfly, *Pieris brassicae* L. An upcoming menace for *brassica* oilseed crops in northern India. Cruciferae Newsletter 25: 83-86.

Lal OP (1975) A Compendium of insect Pest of vegetables in India. Bull Entomol 16: 31-56.

Lebedeva KV, Vendilo NV, Ponomarev VL, Pletnev VA, Mitroshin DB (2002) Use of pheromones and other semiochemicals in integrated production, IOBC wprs Bulletin 25.

Leyrer LR, Monroe ER (1973) Isolation and identification of the scent of the moth *Galleria mellonella*, and reevaluation of its sex pheromone. J. Insect Physiol 19: 2267 - 2271.

Lindgren BS, Borden JH, Pierce AM, Pierce Jr HD, Oehlschlaoer AC, Wong JW (1985) A potential method for simultaneous, semiochemical-based monitoring of *Cryptolestes ferrugineus* and *Tribolium castaneum* (Coleoptera: Cucujidae and Tenebionidae). J Stored Prod Res 21: 83-87.

Linnæus C (1758) Systema naturæ per regna tria naturæ, secundum classes, ordines, genera, species, cum characteribus, differentiis, synonymis, locis. Tomus I. (Ed) decima, Reformata 1-824.

Li Y, Mathews RA (2016) In vivo real-time monitoring of aphrodisiac pheromone release of small white cabbage butterflies (*Pieris rapae*). J Insect Physiol 91-92: 107-112.

Loschiavo SR, Wong J, White NDG, Pierce Jr HD (1986) Field evaluation of a pheromone to detect adults rusty grain beetles, *Cryptolestes ferrugineus* (Coleoptera: Cucujidae), in stored grain. Can Entomol 118(1): 1-8.

Lucas E, Coderre D, Brodeur J (1998) Intraguild predation among aphid predators: characterization and influence of extraguild prey densities. Ecology 79: 1084-1092.

Lu P-F, Qiao H-L, Luo Y-Q (2013) Female sex pheromone blends and male response of the legume pod borer *Maruca vitrata* (Lepidoptera: Crambidae), in two populations of mainland China. Zeitschrift für Naturforschung C 68: 416-427.

Matsumoto K, Nakamuta K, Nakashima T (2007) Mating disruption controls the cherry tree borer, *Synanthedon hector* (Butler) (Lepidoptera: Sesiidae), in a steep orchard of cherry trees. J For Res 12: 34-37.

McDanell R, McLean AEM, Hanley AB, Heaney RK, Fenwick GR (1988) Chemical and biological properties of indole glucosinolates (glucobrassicins): a review. Food Chem Toxicol 26: 59-70.

McElfresh J, Millar J (2001). Geographic Variation in the Pheromone System of the Saturniid Moth *Hemileuca eglanterina*. Ecology 82(12): 3505-3518.

McQueen EW, Morehouse NI (2018) Rapid divergence of wing volatile profiles between subspecies of the butterfly *Pieris rapae* (Lepidoptera: Pieridae). J Chem Ecol 44: 525-533.

Meinwald J, Meinwald YC, Wheeler JW, Eisner T, Brower LP, (1966) Major components in the exocrine secretion of a male butterfly (lycorea). Science 151: 583-585.

Melchini, A., Traka, M. H., Catania, S., Miceli, N., Taviano, M. F., Maimone, P., et al. (2013). Antiproliferative activity of the dietary isothiocyanate erucin, a bioactive compound from cruciferous vegetables, on human prostate cancer cells. Nutr Cancer 65: 132-138.

Mewis I, Appel HM, Hom A, Raina R, Schultz JC (2005) Major Signaling Pathways Modulate Arabidopsis Glucosinolate Accumulation and Response to Both Phloem-Feeding and Chewing Insects. Plant Physiol 138: 1149-1162.

Mewis I, Tokuhisa JG, Schultz JC, Appel HM, Ulrichs C, Gershenzon J (2006) Gene expression and glucosinolate accumulation in *Arabidopsis thaliana* in response to generalist and specialist herbivores of different feeding guilds and the role of defense signaling pathways. Phytochemistry 67: 2450-2462.

Minks AK, Roelofs WL, Ritter FJ, Persoons CJ (1973) Reproductive isolation of two tortricid moth species by different ratios of a two-component sex attractant. Science 180: 1073-1074.

Mizunami M, Yamagata N, Nishino H (2010) Alarm pheromone processing in the ant brain: an evolutionary perspective. Front behav neurosci, 4: 28.

Mondello L, Dugo P, Bade A, Dugo G (1995) Interactive use of linear retention indices, on polar and apolar columns, with a MS-library for reliable identification of complex mixtures. J. Microcolumn Separations 7(6): 581-591.

Monteys S, Quero C, Santa-Cruz MC, Rosell G, Guerrero A (2016) Sexual communication in day-flying Lepidoptera with special reference to castniids or 'butterfly-moths'. Bull Entomol Res 106(4): 421-431.

Moyes CL, Collin HA, Britton G, Raybould AF (2000) Glucosinolates and differential herbivory in wild populations of *Brassica oleracea*. J Chem Ecol 26: 2633-2641.

Mucha-Pelzer T, Mewis I, Ulrichs C (2010) Response of Glucosinolate and Flavonoid Contents and Composition of *Brassica rapa* ssp. chinensis (L.) Hanelt to Silica Formulations Used as Insecticides. J Agric Food Chem 58: 12473-12480.

Mukerji GP (1961) On the biology of "cabbage white", *Pieris brassicae* L. J Zool Soc India 13: 121-127.

Natale D, Pasqualini E (1999) Il controllo di zeuzera e cossus mediante feromoni. Inf Agrar 16:79-83.

Nieberding CM, de Vos H, Schneider VM, Lassance JM, Estramil N, Andersson J, Bang J, Hedenström E, Löfstedt C, Brakefield PM (2008) Male sex pheromone of the butterfly *Bicyclus anynana* towards an evolutionary analysis. PLoS ONE 3(7): 2751.

Nishida R, Schulz S, Kim CS, Fukami H, Kuwahara Y, Honda K, Hayashi N (1996) Male sex pheromone of a gaint Danaine butterfly, *Idea leuconoe*. J Chem Ecol 22(5) 949-972.

Nordlund et al. 1981

Obara Y (1964) Mating behaviour of the cabbage white, *Pieris rapae crucivora* II: the 'mate-refusal' posture of the female. Zool Mag 73: 175-178.

Obara, Y, Hidaka T (1968) Recognition of Female by Male on Basis of Ultra-Violet Reflection in White Cabbage Butterfly *Pieris rapae crucivora* Boisduval. P Jpn Acad 44: 829-832.

Obara Y, Ozawa G, Fukano Y (2008) Geographic variation in ultraviolet reflectance of the wings of the female cabbage butterfly, *Pieris rapae* L. Zool Sci 25(11): 1106-1110.

Ohguchi O, Hidaka T (1988) Mate recognition in two sympatric species of butterflies, *Pieris rapae* and *Pieris melete*. J Ethol 6: 49-53.

Ômura H, Honda K (2009) Behavioral and electroantennographic responsiveness of adult butterflies of six nymphalid species to food derived volatiles. Chemoecology 19(4): 227-234.

Ômura H, Honda K, Hayashi N (1999) Chemical and chromatic bases for preferential visiting by the cabbage butterfly, *Pieris rapae*, to rape flowers. J Chem Ecol 25(8): 1895-1906.

Pare PW, Tumlinson JH (1999) Plant volatiles as a defense against insect herbivores. Plant Physiol 121: 325-331.

Park KC, Ochieng SA, Zhu J, Baker TC (2002) Odor discrimination using insect electroantennogram responses from an insect antennal array. Chem Senses 27(4): 343-352.

Peters A (1996) The natural host range of *Steinernema* and *Heterorhabditis* spp. and their impact on insect populations. Biocontrol Sci Technol 6: 389-402.

Phelan PL, Silk PJ, Northcott CJ, Tan SH, Baker TC (1986) Chemical identification and behavioral characterization of male wing pheromone of *Ephestia elutella* (Pyralidae). J Chem Ecol 12(1) 135-146.

Pliske TE, Eisner T (1969) Sex Pheromone of the Queen Butterfly: Biology. Science 164(3884): 1170-1172.

Pickett JA, Wadhams LJ, Woodcock CM (1997) Developing sustainable pest control from chemical ecology. Agr Ecosyst Environ 64: 149-156.

Pivnick KA, Lamb RJ, Reed D (1992) Response of flea beetles, *Phyllotreta* spp., to mustard oils and nitriles in field trapping experiments. J Chem Ecol 18: 863-873.

Powell JA, Opler PA (2009) Moths of Western North America. University of California Press. http://www.jstor.org/stable/10.1525/j.ctt1pnh89.5.

Prakash D, Gupta C (2012) Glucosinolates: The phytochemicals of nutraceutical importance. J Complement Integr Med 9.

Rask L, Andreasson E, Ekbom B, Eriksson S, Pontoppidan B, Meijer J (2000) Myrosinase: gene family evolution and herbivore defence in Brassicaceae, Plant Mol. Biol. 42: 93-113.

Renwick JAA, Radke CD, Sachdev-Gupta K (1992) Leaf surface chemicals stimulating oviposition by *Pieris rapae* (Lepidoptera: Pieridae) on cabbage. Chemoecology 3: 33-38.

Richards OW (1940) Biology of *Pieris rapae* with special reference to factors controlling abundance. J Anim Ecol 9: 243-288.

Rodman JE, Chew FS (1980) Phytochemical correlates of herbivory in a community of native and naturalized Cruciferae. Biochem Syst Ecol 8: 43-50

Rohr F, Ulrichs C, Mucha-Pelzer T, Mewis I (2006) Variability of aliphatic glucosinolates in Arabido.psis and their influence on insect resistance. Commun Agric Appl Biol Sci 71(2 Pt B): 507-515.

Rohr F (2009) Variabilität aliphatischer Glucosinolate in *Arabidopsis thaliana* Ökotypen und deren Einfluß auf die Wirtspflanzeneignung von zwei folivoren Insektenarten, in: C. Ulrichs, C. Büttner (Eds.), Berliner ökophysiologische und phytomedizinische Schriften 4, Der Andere Verlag, Tönning, Lübeck and Marburg.

Romel KE, Scott-Dupree CD, Caster MH (1992) Qualitative and quantitative analysis of volatiles and pheromone gland extracts collected from *Galleria mellonella* (L.) (Lepidoptera: Pyralidae). J Chem Ecol 18: 1255-1258.

Rothschild M, Schoonhoven LM (1977) Assessment of egg load by *Pieris brassicae* (Lepidoptera: Pieridae). Nature, Lond 266:352-355.

Rosa EA, Hearney RK, Fenwick GR, Portas CA (1997) Glucosinolates in crop plants. Hort Rev 19: 99-215.

Rostas M, Hilker M (2002) Feeding damage by larvae of the mustard leaf beetle deters conspecific females from oviposition and feeding. Entomol Exp Appl 103(3): 267-277.

Rutowski RL (1976) The courtship of *Eurema lisa* (Lepidoptera, Pieridae): A study of visual and chemical communication. Ph.D. Dissertation Cornell University, Ithaca.

Rutowski RL (1977a) Chemical communication in the courtship of small sulfur butterfly *Eurema lisa* (Lepidoptera, Pieridae). J Comp Physiol 115: 75-85.

Rutowski RL (1977b) Use of visual cues in sexual and species discrimination by males of small sulfur butterfly *Eurema lisa* (Lepidoptera, Pieridae). J Comp Physiol 115: 61-74.

Rutowski RL (1980) Male scent-producing structures in *Colias* butterflies –function, localization, and adaptive features. J Chem Ecol 6: 13-26.

Rutowski RL (1991) The evolution of male mate-locating behavior in butterflies. Am Nat 138: 1121-1139.

Sachen JN, Gangwar SK (1980) Vertical distribution of important pest of Cole crop in Meghalaya as influenced by the environment factors. Indian J Entomol 42: 414-421.

Santolamazza-Carbone S, Velasco P, Soengas P, Cartea ME (2014) Bottom-up and top-down herbivore regulation mediated by glucosinolates in *Brassica oleracea* var. acephala. Oecologia 174: 893-907.

Sasaerila Y, Gries R, Gries G, Khaskin G, King S, Takacs S, Hardi (2003) Sex pheromone components of male *Tirathaba mundella* (Lepidoptera: Pyralidae). Chemoecology 13: 89-93.

Schläger S (2015) Identification of variation within sex pheromone blends of various *Maruca vitrata* populations for refining pheromone lures and traps in Asia. Ph.D. Dissertation Humboldt Universität zu Berlin, Berlin.

Schläger S, Ulrichs C, Srinivasan R, Beran F, Bhanu KRM, Mewis I, Schreiner M (2012) Developing pheromone traps and lures for *Maruca vitrata* in Taiwan. Gesunde Pflanze 64 (4): 183-186.

Schneider D (1957) Elektrophysiologische Untersuchungen von Chemo- und Mechanorezeptoren der Antenne des Seidenspinners *Bombyx mori* L. Z Vergl Physiol 40: 8-41.

Schoonhoven LM (1967) Chemoreception of mustard oil glycosides in larvae of *P. brassicae*. Koninkl. Ned. Akad. Wetensch. Proc. Ser. C. 70: 556-568.

Schoonhoven LM (1969) Amino-acid reception in larvae of *P. brassicae* (Lepidoptera). Nature, Lond. 221: 1268.

Schoonhoven LM, van Loon JJA (2002) An inventory of taste in caterpillars: each species itsown key. Acta Zool. Acad. Sci. Hung. 48 (Suppl. 1):215–263.

Schoonhoven LM (1967) Chemoreception of mustard oil glucosides in larvae of Pieris brassicae L. Proc K Ned Akad Wet C 5: 556-568.

Schoonhoven LM (1969) Gustation and foodplant selection in some lepidopterous larvae. Entomol exp appl 12: 555-564.

Schott M, Wehrenfennig C, Gasch T, Duering R-A, Vilcinskas A (2013) A portable gas chromatograph with simultaneous detection by mass spectrometry and electroantennography for the highly sensitive in situ measurement of volatiles. Anal Bioanal Chem 405: 7457-7467.

Schulz S, Boppre M, Vane-Wright RI (1993) Specific mixtures of secretions from male scent organs of African milkweed butterflies (Danainae). Phil Trans R Soc Lond B 342: 161-181.

Schulz S, Nishida R (1996) The pheromone system of the male danaine butterfly, *Idea leuconoe*. Bioorg Med Chem 4: 341-349.

Schulz S, Peram PS, Menke M, Hölting S, Räpke R, Melnik K, Poth D, Mann F, Henrischen, Drezer K (2017) Mass spectrometry of aliphatic macrolides, important semiochemicals or Pheromones. J Nat Prod 80(9): 2572-2582.

Schulz S, Yildizhan S, van Loon JJA (2011) The biosynthesis of hexahydrofarnesylacetone in the butterfly *Pieris brassicae*. J Chem Ecol 37: 360-363.

Shapiro VA (1975) Ecological behavioral aspects of consistence in six crucifers feeding Pierid butterflies in the central Sirra Nevada. Am Midl Nat 424-433.

Shorey HH (1966) The biology of *Trichoplusia ni* (Lepidoptera: Noctuidae). IV. Environmental control of mating. Ann Entomol Soc Am 59(3): 502-506.

Silberglied RE (1984) Visual communication and sexual selection among butterflies. In the biology of butterflies (ed. R. I. Vane-Wright and P. R. Ackery) London: Academic Press 207-223.

Slansky FJ, Scriber JM (1985) Food Consumption and utilization. Kerkut GA, Gilbert LI (Eds) Comprehensive insect physiology biochemistry and pharmacology Oxford, UK Pergamon 4: 87-163.

Smallegange RC, van Loon JJ, Blatt SE, Harvey JA, Agerbirk N, Dicke M (2007) Flower vs. leaf feeding by *Pieris brassicae*: Glucosinolate-rich flower tissues are preferred and sustain higher growth rate. J Chem Ecol 33: 1831-1844.

Sourakov A, Duehl A, Sourakov A (2012) Foraging behaviour of the Blue Morpho and other tropical butterflies: The chemical and electrophysiological basis of olfactory preferences and the role of color. Psyche: Hindawi Publishing Corporation 1-10.

Srinivasan R, Lin MY, Su FC, Yule S, Khumsuwan C, Hien T, Hai VM, Khanh LD, Bhanu KRM (2015) Use of insect pheromones in vegetable pest management: Successes and struggles. In: Chakravarthy AK (ed) New horizons in insect science: Towards sustainable pest management. Springer India 231-237.

Suzuki Y, Nakanishi A, Shima H, Yata O, Saigusa T (1977) Mating behavior in four Japanese species of the genus *Pieris* (Lepidoptera, Pieridae). Kontyu. 45: 300-313.

Textor S, Greshenzon J (2009) Herbivore induction of the Glucosinolate-myrosinase defense system: major trends, biochemical bases and ecological significance. Phytochem Rev 8(1): 149-170.

Tang Y, Zhou C, Chen X, Zheng H (2013) Foraging Behavior of the Dead Leaf Butterfly, *Kallima inachus*. J Insect Sci 13(58) 58.

Tanaka S, Nishida T, Ohsaki N (2007) Sequential rapid adaptation of indigenous parasitoid wasps to the invasive butterfly *Pieris brassicae*. Evolution 61: 1791-1802.

Taylor OR (1973) Reproductive isolation in *Colias eurytheme* and *C. philodice* (Lepidoptera: Pieridae): Use of olfaction in mate selection. Ann entomol Soe Amer 66: 621-626.

Traka M, Mithen R (2009) Glucosinolates, isothiocyanates and human health. Phytochem Rev 8: 269-282.

Traynier RMM, Truscott RJW (1991) Potent natural egg-laying stimulant for cabbage butterfly *Pieris rapae*. J Chem Ecol 17: 1371-1380.

Turlings TCJ, Alborn HT, Loughrin JH, Tumlinson JH (2000) Volicitin, An elicitor of maize volatiles in oral secretion of *Spodoptera Exigua*: Isolation and bioactivity. J Chem Ecol 26(1): 189-202.

Turlings TCJ, Wackers FL, Vet LEM, Lewis WJ, Tumlinson JH (1993) Learning of host location cues by insect parasitoids. In insect learning: Ecological and evolutionary perspectives; Papaj DR, Lewis A, (ed) Chapman and Hall: New York, USA 51-78.

Uechi K, Matsuyama S, Suzuki T (2007) Oviposition attractants for *Plodia interpunctella* (Hübner) (Lepidoptera: Pyralidae) in the volatiles of whole wheat flour. J Stored Prod Res 43: 193-201.

Uehara T, Naka H, Matsuyama S, van Vang L, Ando T, Honda H, (2013) Identification of conjugated pentadecadienals as sex pheromone components of the Sphingid moth, *Dolbina tancrei*. J Chem Ecol 39: 1441-1447.

Van Den Dool H, Kratz PD (1963) A generalization of the retention index system including linear temperature programmed gas-liquid partition chromatography. J Chromatogr A 11: 463-471.

Van Driesche RG, Hoddle M (1997) Should arthropod parasitoids and predators be subject to host range testing when used as biological control agents? Agric Human Val 14: 211-226.

Van Loon JJA, Blaakmeer A, Griepink FC, Van Beek TA, Schoonhoven LM, Degroot Ae (1992) Leaf surface compound from *Brassica oleracea* (Cruciferae) induces oviposition by *Pieris brassicae* (Lepidoptera: Pieridae). Chemoecology 3: 39-44.

Van Nieukerken EJ, Lauri K, Ian JK, Kristensen NP, Lees DC, Minet J, Mitter C, Mutanen M, Regier JC, Simonsen T, Wahlberg N, Yen S, Zahiri R, Adamski D, Baixeras J, Bartsch D, Bengtsson BA, Brown JW, Bucheli S, Zwick A (2011). Order Lepidoptera Linnaeus, 1758. In: Zhang Z-Q (ed) Animal biodiversity: An outline of higher classification and survey of taxonomic richness.

Verheggen F, Haubruge E, Mescher M (2010) Alarm pheromones - Chemical signaling in response to danger. Vitam Horm 83: 215-39.

Vial KM, Kok LT, McAvoy TJ (1991) Cultivar preferences of lepidopterous pests of broccoli. Crop Prot 10: 199-204.

Weiss I, Rössler T, Hofferberth J, Brummer M, Ruther J, Stökl J (2013) A nonspecific defensive compound evolves into a competition avoidance cue and a female sex pheromone. Nat Commun 4: 2767.

Weiss I, Hofferberth J, Ruther J, Stökl J (2015). Varying importance of cuticular hydrocarbons and iridoids in the species-specific mate recognition pheromones of three closely related *Leptopilina* species. Front Ecol Evol 3(19): 1-12.

West LG, Meyer KA, Balch BA, Rossi FJ, Schultz MR, Haas GW (2004) Glucoraphanin and 4-Hydroxyglucobrassicin contents in seeds of 59 cultivars of broccoli, raab, kohlrabi, radish, cauliflower, brussels sprouts, kale, and cabbage. J Agric Food Chem 52(4): 916-926.

Wiklund C, Forsberg J (1985) Courtship and male discrimination between virgin and mated females in the orange tip butterfly, *Anthocharis cardamines*. Anim Behav 34:328-332.

Willstätter R, Stoll A (1913) Untersuchungen über Chlorophyll. Verlag von Julius Springer.

Witzgall P, Kirsch P, Cork A, (2010) Sex pheromones and their impact on pest management. J Chem Ecol 36: 80-100.

Wong JW, Verigin V, Oelschlager AC, Borden JH, Pierce AM, Pierce HD Jr, Chong L, (1983) Isolation and identification of two macrolide pheromones from the frass of *Cryptolestes ferrugineus* (Coleoptera: Cucujidae). J Chem Ecol 9(4): 451.

Wyatt T (2010) Pheromones and signature mixtures: Defining species-wide signals and variable cues for identity in both invertebrates and vertebrates. J Comp Physiol A https://doi.org/10.1007/s00359-010-0564-y.

Yaginuma K (1984) Cherry tree borer. Experimental method of pheromones. JPPA 2: 116-120 (in Japanese).

Yildizhan S, van Loon J, Sramkova A, Ayasse M, Arsene C, ten Broeke C, Schulz S (2009) Aphrodisiac pheromones from the wings of the small cabbage white and large cabbage white butterflies, *Pieris rapae* and *Pieris brassicae*. Chem Bio Chem 10: 1666-1677.

Yoshida A, Noda A, Yamana A, Numata H, (2000) Arrangement of scent scales in the male wings of small white cabbage butterfly (Lepidoptera: Pieridae). Entomol Sci 3(2): 345-349.

Younas M, Naeem, M, Raqib A, Masud S (2004) Population dynamics of Cabbage butterfly (*Pieris brassicae*) and cabbage aphids (*Brevicoryne brassicae*) on five cultivars of cauliflower at Peshawar. Asian J Plant Sci 3: 391-393.

Zimmermann Y, Ramirez SR, Eltz T (2009) Chemical niche differentiation among sympatric species of orchid bees. Ecology 90: 2994-3008.

8 Summary

Chemical ecology leads to study of chemicals specifically released from one specie and can affect the same or other species. Two *Pieris* species have been selected for this study i.e., *Pieris rapae* and *Pieris brassicae*, that are causing damages to the agricultural *brassica* crops throughout the world. Some specific compounds that are considered as aphrodisiac pheromones are produced and released from the wings of butterflies. They play an important role for a successful courtship. They are present in different concentrations in both the conspecifics. In this study the volatile compounds from wings of male and female butterflies have been isolated and quantified.

A comparative study among four different populations of *P. rapae* was carried out. Two Asian populations, *P. rapae crucivora* (Taiwan, Vietnam) were compared with two European populations, the *P. rapae rapae* (Germany, Netherlands). The total pheromone levels in the european populations were lower than amounts present in asian populations. In the populations from Europe, ferrulactone was one of the major constituents of male-specific pheromones, while E-phytol was the major pheromone compound in the asian populations from Taiwan and Vietnam. As a result of this study, we can conclude that both the sub-species *P. rapae crucivora* and *P. rapae rapae* showed a clear variation in pheromone profile between them, but within the sub-species, the different populations did not show so large differences. Therefore, the populations with geographical difference could not be considered identical as they showed quantitative differences in pheromone profiles. Furthermore, it was found via EAG analysis that female antennae respond against specific concentrations of ferrulactone and hexahydrofarnesylacetone and they could be used as pheromone lures for both the subspecies to carried out field bioassays to test the mating disruption strategy.

Another experiment was carried out using *P. brassicae* German population. Brassicalactone was found to be a male-specific compound. Isophytol and hexahydrofarnesylacetone were found in the wings extracts of both conspecifics with no difference in quantities. E-phytol was found in males in ten-folds more than in females. A small bioassay was also carried out using different ratios of synthetic E-Phytol, isophytol and hexahydrofarnesylacetone. As a result, we can conclude that a specific dose of these three compounds of synthetic origin can also play an important role in attraction of female butterflies to the males. These doses can be further used in the field trails in future studies and this study of wings compounds could be advantageous for the integrated pest management studies.

Another aspect of chemical ecology leads to the study of Glucosinolates. In this part of research, a host plant screening was performed using two specialist herbivores, the European *P. rapae* and *P. brassicae* and the induction of glucosinolates in *Brassica* varieties upon specialist herbivory was also investigated. All five *Brassica* varieties were found to be good host plants for *P. rapae* larvae, while *P. brassicae* larvae showed the best growth only on Marrow-stem Kale. Our GS analysis indicates that *P. rapae* feeding elicited the increase in aliphatic GS content only in Marrow-stem Kale, whereas *P. brassicae* herbivory showed increase in Marrow-stem as well as in Savoy cabbage. The indolyl GS levels in all *Brassica* varieties were increased to several-folds upon feeding of both specialist pests. This study would be advantageous to choose host plant varieties for these two herbivores.

9 Zusammenfassung

Die chemische Ökologie beschäftigte sich mit der Funktion von Botenstoffen in den Wechselbeziehungen von Organismen. In der vorliegenden Arbeit erfolgten die Untersuchungen zu diesen Botenstoffen zweier *Pieris*-Arten. *Pieris rapae* und *Pieris brassicae*, zwei kosmopolitische Schädlinge an landwirtschaftlichen Kulturen der Gattung Brassicaceae. Einige der Stoffe welche in den Flügeln der Schmetterlinge produziert werden funktionieren als Aphrodisiakum. Sie spielen eine zentrale Rolle für eine erfolgreiche Paarung und werden in beiden Arten in unterschiedlichen Mengen produziert. In der vorliegenden Studie wurden die verdampfbaren Substanzen aus den Flügeln männlicher und weiblicher Schmetterlinge isoliert und quantifiziert.

Es wurde eine Vergleichsstudie mit vier verschiedenen Populationen von *P. rapae*, zwei asiatischen Populationen (*P. rapae crucivora* aus Taiwan & Vietnam) und zwei europäischen Populationen (*P. rapae rapae* aus Deutschland & Niederlande) durchgeführt. Die Duftstoffkonzentration in den europäischen Populationen war niedriger als die in den asiatischen Populationen. In den Populationen aus Europa war Ferrulacton eines der Hauptbestandteile des vom Männchen produzierten Pheromons, während E-Phytol ein Hauptbestandteil in den asiatischen Populationen war.

Im Ergebnis dieser Studie kann geschlussfolgert werden, dass die Subspezies *P. rapae crucivora* als auch *P. rapae rapae* eine deutliche Variation des Pheromonprofils zeigten. Im Gegensatz dazu gab es innerhalb der Unterarten keine großen Unterschiede. Darüber hinaus wurde in Elektroantennogramm-Studien festgestellt, dass weibliche Antennen auf Ferrulacton und Hexahydrofarnesylaceton reagieren und als Pheromonköder für beide Subspezies verwendet werden können.

In einem weiteren Experiment mit der deutschen Population von *P. brassicae* konnte gezeigt werden, dass Brassicalacton ein männlich-spezifisches Pheromon ist. Isophytol und Hexahydrofarnesylaceton wurden in den Flügelextrakten beider Geschlechter ohne Mengenunterschied identifiziert. E-Phytol wurde bei den Männchen in zehnfacher Konzentration nachgewiesen, als in den weiblichen Falter.

Ein weiterer Aspekt behandelte die Induktion von Sekundärmetaboliten (Glucosinolaten) in den Wirtspflanzen. In diesem Teil der Arbeit wurde ein Wirtspflanzen-Screening mit europäischen *P. rapae* und *P. brassicae*, durchgeführt. Alle fünf getesteten *Brassica*-Sorten (Rosenkohl (*Brassica oleracea* var. gemmifera DC. 'Igor'), Grünkohl (*Brassica oleracea* var. sabellica L. 'Halbhoher Grüner Krauser'), Markstammkohl (*Brassica oleracea* var. medullosa Thell. 'Grüner Ring'), Chinakohl (*Brassica rapa* ssp. pekinensis 'Kasumi F1'), und Wirsing (*Brassica oleracea* var. sabauda L. "Vorbote/Hilmar)) erwiesen sich als geeignete Wirtspflanzen für *P. rapae*-Larven, während sich *P. brassicae*-Larven nur auf Markstammkohl sehr gut entwickelten. GS-Analyse zeigten, dass *P. rapae* nur bei Markstammkohl eine Induktion der aliphatischen GS bewirkte, während *P. brassicae* sowohl bei Markstamm als auch bei Wirsing zu einer Induktion führte. Die Indolyl-GS-Level in allen *Brassica* Sorten wurden bei Fütterung beider Schädlinge auf ein Mehrfaches erhöht.

Zusammenfassung

10 Acknowledgments

All praises be to Allah Almighty who showered upon me His unlimited blessings. And peace be upon His messengers, the last of whom was Prophet Muhammad (SAW), who brought the light of knowledge and wisdom to the mankind.

I am heartily indebted to my supervisor, Prof. Dr. Dr. Christian Ulrichs, whose motivation, guidance, and support from the initial to the final level enabled me to develop an understanding of the subject.

I am very grateful to Dr. Inga Mewis for her kind personal support throughout all my studies. She guided me as a good mentor at each and every step. I am also very thankful to the people working in AVRDC, the World Vegetable Centre, Taiwan and Vietnam for sending the insects rearing for me.

I am deeply indebted to my lab-fellow Dr. Nadja Förster whose presence and experience in the same lab was a great help for me. I also thank Susanne Meier for her really nice and cooperative working habits.

My special thanks also go to Dr. Cornelia Lehmann and Dr. Stefanie Schläger for their precious time and assistance, especially at the very beginning of my studies.

I thank all my colleagues, Stefan Irrgang, Armin Blievernicht, Anje Schmidt, YoungJong Han, Vanessa Hörmann, Olivia Naa Ayorkor Tetteh, Julia Eckardt, Jan-Christoph Gloger for delightful times and creating an energetic working atmosphere.

I acknowledge deeply the whole group of ecological chemistry, plant analysis and stored product protection, Julius-Kühn Institute especially, Dr. Cornel Adler, Dr. Tina Gasch, Dr. Christina Müller, Sarah Awatar-Salendo, Dr. Christoph Böttcher, Dr. Benjamin Fürstenau, Rene Grunewald, Tobias and Heidi.

I am heartily thankful to my close friends Bindia, Fozia, Rahila, Hina, Fatima and Iqra for their unconditional love and support. During my stay in Berlin, I enjoyed their company a lot. Be it the excellent food and interesting discussions; I enjoyed every bit with them. They are just like my family members, and I wish them a happy and prosperous life.

I am deeply indebted to my family, Shahid, Eshal, Manal & Nawal for their support and better understanding. They, being a real motivation of my life, inspired me to fulfill this work. I am extremely delighted to thank my Mama, Papa, Mohni, Imran, Ema and Fatima for their love and support. I am grateful to my in-law family members (Javid, Asif, Zahid, Mamoon Tariq and Baji Uzma) for their encouraging support.

11 Selbstständigkeitserklärung

Die vorliegende Arbeit wurde von mir selbst, und nur unter Verwendung der angegebenen Quellen und Hilfsmittel erstellt.

Diese Arbeit wurde nicht zuvor bei einer anderen Hochschule als Dissertation eingereicht. Die geltende Promotionsordnung (Nr.24/2005) der Lebenswissenschaftlichen Fakultät der Humboldt-Universität zu Berlin ist mir bekannt.

Ich erkläre weiterhin, dass ich noch keinen Doktorgrad erlangt oder zu erlangen versucht habe.

Maliha Gul Aftab

Maliha Aftab

Berlin, 01.02.2021

In der Reihe *Berliner ökophysiologische und phytomedizinische Schriften* sind bisher erschienen:

Band 01: Mohammad Mahir Uddin (2009)
Chemical ecology of mustard leaf beetle Phaedon cochleariae (F.).
ISBN 978-3-89959-848-3.

Band 02: Ilir Morina (2009)
Entwicklung von Verfahren zur Rekultivierung der Aschedeponie des Braunkohlekraftwerks in Prishtina (Kosovo).
ISBN 978-3-89959-872-8.

Band 03: Melanie Wiesner (2009)
Veränderungen gesundheitsrelevanter Inhaltsstoffe in *Parthenium hysterophorus* L. in Abhängigkeit von der Pflanzengröße und Klimafaktoren.
ISBN 978-3-89959-880-3.

Band 04: Fransika Rohr (2009)
Variabilität aliphatischer Glucosinolate in *Arabidopsis thaliana*-Ökotypen und deren Einfluss auf die Wirtspflanzeneignung von zwei folivoren Insektenarten.
ISBN 978-3-89959-884-9.

Band 05: Jutta Buchhop (2009)
Characterization of phylogenetically diverse CLRV-isolates by RFLP and research into identification of two isometric viruses.
ISBN 978-3-89959-929-9.

Band 06: Nora Koim (2010)
Urban sprawl, land cover change and forest fragmentation – Case study Pereira, Colombia.
ISBN 978-3-89959-955-8.

Band 07: Nadja Förster (2010)
Eignung unterschiedlicher salicylathaltiger *Salix*-Klone für die Arzneimittelindustrie.
ISBN 978-3-89959-964-0.

Band 08: Jana Gentkow (2010)
Cherry leaf roll virus (CLRV): Charakterisierung ausgewählter Virusisolate unter besonderer Berücksichtigung des viralen Hüllproteins.
ISBN 978-3-89959-976-3.

Band 09: Ahmad Fakhro (2010)
Interaction of Pepino mosaic virus (PepMV) and fungal root endophytes with tomato hosts (*Lycopersicum esculentum* Mill.).
ISBN 978-3-89959-995-4.

Band 10: Stefan Irrgang (2010)
Mikro- und makroskopische Untersuchungen an Veredelungsstellen von Straßenbäumen im Hinblick auf die Beeinflussung ihrer Bruchsicherheit.
ISBN 978-3-89959-998-5.

Band 11: Julia Jahnke (2010)
Guerilla Gardening anhand von Beispielen in New York, London und Berlin.
ISBN 978-3-86247-001-3.

Band 12: Astrid Karoline Günther (2010)
Analysen zur Intensität der Pflanzenschutzmittel-Anwendung und Aufklärung ihrer Einflussfaktoren in ausgewählten Ackerbaubetrieben.
ISBN 978-3-86247-005-1.

Band 13: Milena A. Dimova (2010)
Untersuchungen zur Epidemiologie von *Pythium aphanidermatum* in Abhängigkeit von den Umgebungsbedingungen bei der Gewächshausgurke (*Cucumis sativus* L.).
ISBN 978-3-86247-033-4.

Band 14: Claudia Patricia Pérez-Rodríguez (2010)
Physiologische Veränderungen in Früchten der Solanaceaengewächse in Abhängigkeit von physikalischen Elicitoren während der Produktion und nach der Ernte.
ISBN 978-3-86247-066-2.

Band 15: Charles Adarkwah (2010)
Integrated management of the stored-product pest insects *Corcyra cephalonica*, *Cadra cautella*, *Sitophilus zeamais* and *Tribolium castaneum* by use of the parasitic wasps *Habrobracon hebetor*, *Venturia canescens*, *Lariophagus distinguendus* and neem seed oil.
ISBN 978-3-86247-077-8.

Band 16: Christoph von Studzinski (2010)
Angewandte Methoden der xenovegetativen Vermehrung.
ISBN 978-3-86247-088-4.

Band 17: Tanja Mucha-Pelzer (2011)
Amorphe Silikate – Möglichkeiten des Einsatzes im Gartenbau zur physikalischen Schädlingsbekämpfung.
ISBN 978- 3-86247-106-5.

Band 18: Diego Miranda (2011)
Effect of salt stress on physiological parameters of cape gooseberry, *Physalis peruviana* L.
ISBN 978- 3-86247-119-5

Band 19: Franziska Beran (2011)
Host preference and aggregation behavior of the striped flea beetle, *Phyllotreta striolata*.
ISBN 978- 3-86247-188-1

Band 20: Mohammed Abul Monjur Khan (2011)
Induced biochemical changes and gene expression in *Brassica oleracea* and *Arabidopsis thaliana* by drought stress and its consequences on resistance to aphids.
ISBN 978- 3-86247-203-1.

Band 21: Sandra Lerche (2012)
Untersuchungen zur Anwendung, Praxiseinführung und molekularen Identifizierung von Stamm V24 des entomopathogenen Pilzes *Lecanicillium muscarium* (Petch) Zare & W. Gams.
ISBN 978- 3-86247-248-2.

Band 22: Carsten Richter (2012)
Entwicklung und Überprüfung eines gasdichten Küvettensystems für Experimente unter hochgradig kontrollierten Bedingungen mit Gaswechselmessungen.
ISBN 978- 3-86247-271-0.

Band 23: Aksana Grineva (2012)
Influence of the two stored grain pest insects *Sitophilus granarius* and *Oryzaephilus surinamensis* on temperature, relative humidity, moisture content, and mould growth in stored triticale.
ISBN 978- 3-86247-279-6.

Band 24: Carmen Büttner & Christian Ulrichs (2012)
Aktuelle Themen in Landwirtschaft und Gartenbau am Beispiel von Südtirol.
ISBN 978- 3-86247-279-6.

Band 25: Juliane Langer (2012)
Molecular and epidemiological characterisation of Cherry leaf roll virus (CLRV).
ISBN 978- 3-86247-279-6.

Band 26: Franziska Rohr-Doucet (2012)
AOP-Variabilität in *Arabidopsis thaliana*-Kreuzungslinien – Auswirkungen auf die Resistenz gegenüber verschieden spezialisierten Lepidopteren-Arten.
ISBN 978- 3-86247-329-8.

Band 27: Vanessa Hörmann (2012)
Lignin als biologische Barriere gegen Schimmelpize in Innenräumen.
ISBN 978- 3-86247-330-4.

Band 28: Jacqueline Kurth (2013)
Auswirkungen verschiedener Düngerzusammensetzungen auf den Ertrag bei Schnittrosen unter Berücksichtigung des Anbauverfahrens.
ISBN 978- 3-86247-336-6.

Band 29: Juliane Langer, Carmen Büttner & Christian Ulrichs (2014)
Kolumbien – klimatische und politische Voraussetzungen für eine landwirtschaftliche Produktion.
ISBN 978- 3-86247-430-1.

Band 30: Heike Luisa Dieckmann (2014)
Detection of the European mountain ash ringspot associated virus (EMARaV) in Sorbus aucuparia L. in several European contries.
ISBN 978- 3-86247-441-7.

Band 31: Rima Marion Baag (2014)
Analyse von trans-Resveratrol in historischen Rebsorten der Weinanbaugebiete Sachsen und Saale-Unstrut.
ISBN 978- 3-86247-488-2.

Band 32: Ayesha Rahmann (2014)
Study of the protective effects of nano-structured silica and plant derived biomolecules on nuclear polyhedrosis virus affected silkworm larvae at the behavioral and molecular level.
ISBN 978- 3-86247-495-0.

Band 33: Bettina Gramberg (2015)
Weiterentwicklung eines elektrochemischen Biosensors zum Nachweis von Pflanzenviren und Insektiziden.
ISBN 978- 3-86247-512-4.

Band 34: Wilhelm van Husen (2015)
Artspezifische Aufnahme und Verteilung von Cadmium bei indigenen afrikanischen Gemüsearten und daraus abzuleitende Ernährungsempfehlungen.
ISBN 978- 3-86247-523-0.

Band 35: Jenny Roßbach (2015)
European mountain ash ringspot-associated viras (EMARaV): diversity and geographic distribution in Europe.
ISBN 978- 3-86247-547-6.

Band 36: Silke Steinmöller (2015)
Risikominderung der Verbreitung von Quarantäneschadorganismen der Kartoffel durch hygienisierende Maßnahmen.
ISBN 978- 3-86247-550-6.

Band 37: Christin Siewert (2016)
Genomic and functional analysis of species within the Acholeplasmataceae – Phytoplasmas and Acholeplasmas.
ISBN 978- 3-86247-579-7.

Band 38: Angela Köhler (2016)
Untersuchungen zur Phenolglycosidkonzentration ausgewählter intra- und interspezifischer Kreuzungen salicinreicher Biomasseweiden.
ISBN 978- 3-86247-581-0.

Band 39: Nicolas Meyer (2016)
Vergleichende ökophysiologische Untersuchung verschiedener Baumarten zur Verwendung als Straßenbegleitgrün in Berlin.
ISBN 978- 3-86247-586-5.

Band 40: Stefanie Schläger (2017)
Identification of variation within sex pheromone blends of various Maruca vitrata populations for refining pheromone lures and traps in Asia.
ISBN 978- 3-7369-9570-3.

Band 41: Elisha Otieno Gogo (2017)
Pre- and postharvest treatments for the quality assurance of African indigenous leafy vegetables.
ISBN 978-3-7369-9650-2.

Band 42: Luise Dierker (2017)
Interaktion des RNA2-kodierten Transportproteins (MP) des *Cherry leaf roll virus* (CLRV) mit dem viralen Hüllprotein (CP) und pflanzlichen Wirtsfaktoren
ISBN 978-3-7369-9670-0.

Band 43: Nadja Förster (2017)
Antikarzinogenes Potential ausgewählter Glucosinolate von *Moringa oleifera*
ISBN 978-3-7369-9704-2.

Band 44: Steffen Pallarz (2018)
Data driven classification of host-plant response (virus-plant)
ISBN 978-3-7369-9731-8.

Band 45: Vanessa Hörmann (2018)
Biofiltration of indoor pollutants by ornamental plants
ISBN 978-3-7369-9815-5.

Band 46: Allan Ndua Mweke (2018)
Development of entomopathogenic fungi as biopesticides for the management of Cowpea Aphid, *Aphis craccivora* Koch
ISBN 978-3-7369-9908-4.

Band 47: Isabella Linda Bisutti (2019)
Biological agents formulation and mode of application against strawberry diseases
ISBN 978-3-7369-7032-8.

Band 48: Judith Henze (2019)
Innovation in Agriculture: The Potential, Challenges and Adoption and Diffusion of Aquaponics and Agricultural Mobile Phone Application in Kenya
ISBN 978-3-7369-7133-2.

www.ingramcontent.com/pod-product-compliance
Ingram Content Group UK Ltd.
Pitfield, Milton Keynes, MK11 3LW, UK
UKHW021959190726
13853UKWH00004B/1628

9 783736 974982